THE HERNDON CLIMB

THE HERNDON CLIMB

A HISTORY OF THE UNITED STATES NAVAL ACADEMY'S GREATEST TRADITION

REAR ADM. JAMES R. McNEAL, SC, USN (RET.)
AND SCOTT TOMASHESKI

Foreword by Gary Kinder

Naval Institute Press
Annapolis, Maryland

Naval Institute Press
291 Wood Road
Annapolis, MD 21402

Library of Congress Cataloging-in-Publication Data
Names: McNeal, James R., author. | Tomasheski, Scott, author.
Title: The Herndon Climb : a history of the United States Naval Academy's greatest tradition / Rear Adm. James R. McNeal, SC, USN (Ret.), and Scott Tomasheski.
Other titles: History of the United States Naval Academy's greatest tradition
Description: Annapolis, MD : Naval Institute Press, [2020] | Includes bibliographical references.
Identifiers: LCCN 2020013689 (print) | LCCN 2020013690 (ebook) | ISBN 9781682474389 (paperback) | ISBN 9781682475522 (epub) | ISBN 9781682475522 (pdf)
Subjects: LCSH: United States Naval Academy—Student life. | Herndon Monument (Annapolis, Md.) | Herndon, William Lewis, 1813–1857—Monuments—Maryland—Annapolis. | Midshipmen—United States—Interviews. | Pyramids (Gymnastics)—United States.
Classification: LCC V415.P1 M35 2020 (print) | LCC V415.P1 (ebook) | DDC 359.0071/173—dc23
LC record available at https://lccn.loc.gov/2020013689
LC ebook record available at https://lccn.loc.gov/2020013690

♾ Print editions meet the requirements of ANSI/NISO z39.48-1992 (Permanence of Paper).
Printed in the United States of America.

28 27 26 25 24 23 22 21 20 9 8 7 6 5 4 3 2 1
First printing

CONTENTS

FOREWORD

IT HARDLY SEEMS A FAIR FIGHT: twelve hundred buff plebes against one skinny obelisk. And the goal is simple: Lift a plebe's cap—a "Dixie Cup"—off the pointy top and crown the obelisk with a peaked officer's "cover." So why does it take an hour or two, and sometimes four? And an even bigger question: Why at all?

It is May and it is hot. There is little shade, but squeezed into whatever shadows they can find, thousands of spectators unpack picnics. Many have watched the Herndon Climb every year for decades. A few have even participated. Next to graduation, "Herndon" is the most significant experience in Academy life.

To begin the event, a small brass cannon fires a shotgun shell, and while the plebes race toward the monument, let's look at the bronze plaque at its base, which the plebes have never seen. Until the Herndon Climb, they have not been allowed to approach the grass or walk on the paths that surround the monument. And the plaque is too small to read from even a few feet away. Had they been allowed closer, the plebes would have seen the name "HERNDON" in raised letters on the back side, and they could have read the one sentence cast on the plaque in 1860: "Forgetful of self, in his death he added a new glory to the annals of the sea."

That is not the full sentence, mind you, but that's all that was cast on the plaque. It will do for now.

A more immediate concern: Do the plebes running toward the monument know that Cdr. William Lewis Herndon is the only officer in the United States Navy ever to be honored with a monument in time of peace? And do they realize the irony that the Herndon Call they are about to answer differs only in degree from the Herndon Call once answered by five hundred steamship passengers returning from the Gold Rush in 1857?

The sentence abbreviated in the casting originally read: "Forgetful of self, *mindful of others, his life was beautiful to the last*; and in his death he has added a new glory to the annals of the sea." The words were part of a eulogy written by the father of modern oceanography, Matthew Fontaine Maury. When Herndon went down with his ship after battling a hurricane for four days, Maury wanted the world to know that a man had once stood flatfooted on the hurricane deck of a sinking ship carrying the souls of six hundred people and lived an oath he had taken on a calm, sunny day years before: that regardless of danger, he would never abandon his ship. As Maury knew, many had taken the oath, but few had lived it. Just before he wrote the final sentence above, Maury had written, "A cry arose from the sea, but not from his lips. The waves had closed about him, and the curtain of night was drawn over one of the most sublime moral spectacles that the sea ever saw."

Maury probably did not know that Herndon had taken the oath a second time, too, during a genteel dinner at the captain's table as the SS *Central America* steamed toward the edge of the storm. He was dining with several guests when convivial conversation turned to a topic popular at the time: shipwrecks. Three years earlier, the crew of a sinking steamer had abandoned their ship to rescue themselves, leaving the passengers to perish. One of the dinner guests later recalled Herndon's charming segue from sinking ships to more pleasant topics: "Well, I'll never survive my ship," he told his guests. "If she goes down, I go under her keel."

In the early 1900s, the ritual of climbing "Herndon" began with plebes snake dancing around the monument, celebrating having survived a grueling first year at the Academy. The dance would have been the first time they had been allowed on the monument's surrounding lawn and the adjacent path, "Lover's Lane." Eventually, the ritual dance would end with one plebe climbing the obelisk, which officially marked the end of their "plebe" status and provoked hundreds of voices chanting "Plebes no more!" Later, an officer's "cover" would be placed at the top for the climber to retrieve, a symbolic touch; and later still, another symbolic touch: a plebe's cap, or "Dixie Cup," would be placed at the top for the climber to replace with the officer's cover. As McNeal and Tomasheski write, "The exchange of caps at the apex further cemented the profound meaning of the Climb as an important step in the life of every midshipman." With that, the Herndon Climb became "a ritual in Academy culture second only to graduation." And that was the last change to the Herndon Climb. Except for one that greatly evened the odds.

Right now, the first plebes have reached the monument and have begun forming a human scaffold around the base, linked arm-in-arm, like a rugby scrum, on top of a rugby scrum, on top of a rugby scrum, with the monument in the middle. And on top of that last scrum a few "monkeys," the lightest, scrappiest, most agile of the soon-to-be-plebes-no-more, will inch their way toward the top, each balancing a cover to replace the Dixie Cup.

To them all, "Herndon" is the name of a long-forgotten someone who did something once upon a time, but few know what or when. For them, "Herndon" has become an event: a rite of passage, a line to cross from being an absurdity—less than *nothing*—to being *persons*; from being *plebes* to being *plebes no more*. To be one, but become the other, they must *do* Herndon.

At a height of twenty-one feet, the monument is hardly imposing. The first officially timed Herndon in 1962 was over in twelve minutes. But over the years, the great leveler "grease" became a tradition, and the time slowed from minutes to hours as the plebes tried to climb twenty-one feet of granite slicked with fifty pounds of lard. And the plebes' only resource, as the authors explain, is "the combined strength of their own bodies" and the ingenuity and teamwork of classmates. And that's where the Herndon Call of today ironically begins to overlay the Herndon Call of 1857.

As seawater pouring into the hold silenced the *Central America*'s steam engines on the third day, Captain Herndon appeared in the dining salon, where the passengers huddled, and called for volunteers to keep the three-hundred-foot steamer afloat like they would a rowboat: by bailing. With passengers hardened by years in the gold fields and packing weapons, mutiny was common, but five hundred men obeyed Herndon, because he did nothing to save himself. From first class to steerage, from miners to carpenters to vaudevillians, they formed three lines from the hold almost four stories up to the deck and passed buckets, basins, and pitchers filled with water throughout the night and into the next day, thirty-six hours, some passing out from exhaustion. Herndon seemed at once in all parts of the ship, cheering the men beyond what they thought they could do. "Work on, m'boys," he would say. "We have hope yet."

Though few passengers had family with them, their selfless teamwork in answering Herndon's call bought just enough hours to transfer all sixty women and children and a hundred men to another ship that appeared miraculously out of the storm. To this day, that rescue two hundred miles at sea is one of the finest hours in American history.

Each year in May, several hundred plebes answer the other Herndon Call. No one's life is at stake, but they can accomplish the Climb only by building a broad human scaffold nearly three stories high with nothing to hold on to but each other, and with their granite nemesis and themselves slippery with lard. One graduate who placed the cover in his plebe year told the authors, "You get a little disappointed when you all crumble down the greasy thing and start going again, but the feeling you have when it's over is like no other."

A few superintendents have tried to tame Herndon, to make it less difficult, less dangerous, but for what are our finest military women and men preparing if not the difficult and the dangerous? The superintendent in 2010 outlawed using grease, but the "scaler" scrambled to the top in two minutes and five seconds. Not much of a challenge and little need for teamwork. The following year, a new superintendent, more concerned about instilling "spirit and camaraderie" within the plebe class and between that class and the classes before who had *done* Herndon, ordered the grease back on the granite. Exhausted but gratified, the plebes that year took two hours and forty-one minutes to place the cover on top.

One commander acknowledged there will always be strains and sprains and cuts and bruises suffered in the Herndon Climb. "But we also don't want to develop a fighting force that we're unable to send out into the desert, send to the front lines, because we pampered them. . . . We've got to have people that are used to doing this kind of thing. [The Climb] is that same kind of teamwork building, even though it isn't combat. It's about the mental fortitude, accepting challenges and getting out of your comfort zone."

"Herndon" has grown beyond a monument, even beyond an event: It has become a metaphor for life in the military—selflessness, perseverance, and teamwork—the ultimate test of a mass of women and men figuring out how to work together toward a common goal, no one caring who gets the glory. As the authors write, "The Climb is generally viewed as an exercise in cooperation and team building as much as it is a grueling physical challenge for those who throw themselves into the crowd, putting their bodies on the line, and pushing the limits of their strength and endurance."

Watch a Herndon Climb on YouTube. It will make you groan and gasp at the near misses and cheer at the small and ultimate triumphs. As you watch, know that much of Herndon's charm was his self-mocking humor: He loved to regale an audience with punch lines that underscored that the joke was on him. Were he to witness the plebes *doing* "Herndon" today, I am certain he would laugh the longest and cheer the loudest. I hope that none of these plebes will be tested as Herndon was, and that if they are, they will remember the example he set and the legacy he left behind.

I also hope *The Herndon Climb* reaches a readership far beyond the Naval Academy. It deserves such a following. Because Admiral McNeal walked the Herndon walk at the end of his own plebe year—a body in the bottom scrum—his experience informs the firsthand accounts he and Tomasheski teased out of others who have been there and *done* "Herndon." Even if you didn't attend the Academy and don't know anyone who did, the personal stories of victory and sacrifice on the Climb and later on duty inspire. When so many of our leaders show little regard for integrity, selflessness, and honor, we need to be reminded of real heroes, like these young women and men and like William Lewis Herndon.

—Gary Kinder, author of the *New York Times* bestseller *Ship of Gold in the Deep Blue Sea: The History and Discovery of the World's Richest Shipwreck*

PREFACE

IT STANDS SILENTLY, an unyielding sentry, one of the oldest monuments on a campus that is rich in history and not poor in monuments.

It predates nearly every other structure on the grounds of the United States Naval Academy. It survived the Civil War and did its part in the battle on the home front in every other conflict since then, any time the Navy has gone into action. It stands ready, to this day, bearing no scars and no signs of weakening despite its extraordinary length of distinguished service.

Herndon.

The locals refer to it by that simple one-word designation, and they all know it. Any sort of official title beyond that has largely been forgotten, buried in dusty piles of old campus maps and well-thumbed but long-retired copies of midshipman handbooks. Civilians and tourists immediately identify themselves as such if they should refer to it as "the Herndon Monument" or inquire of anyone within a forty-mile radius of Annapolis where it might be found.

The name of Herndon hardly towers over the pages of American history and Navy lore in the manner of a Jones or Bancroft or Nimitz or McCain. Outside of Academy circles and the city of

Annapolis, it is not a household name, even in many Navy homes. But Herndon, the monument, stands because of the bravery and sacrifice of Herndon, the man.

William Lewis Herndon was a man who earned his monument, as surely as the others. He commanded a warship, the USS *Iris*, assisting in the blockade of the coast of Mexico during the Mexican-American War; but his Navy career is primarily distinguished by his unique contributions as an explorer, scientist, writer, and extraordinarily able seaman. He virtually defined the concept of honor, in a bold sacrifice, while serving as captain of a merchant vessel that went to the bottom of the sea, a victim of a merciless Atlantic storm.

Herndon, the monument, stands in front of the Chapel and its spectacular, iconic dome and adjacent to the famous bandstand named for George Zimmerman, the Academy's first and longest-serving bandmaster. The monument itself is a simple granite obelisk, much shorter than the Washington Monument, but similar in shape and design. Two sides are marked, in raised lettering, with "HERNDON" on one side and "September 12, 1857" on the opposite, stark and modest reminders of Commander Herndon and the date of his loss. The other two sides of the monument are perfectly smooth.

Twenty-one feet of sheer-sided granite might be as lost as a needle among the Academy's haystack of offerings of architectural delight, vastness, artistic whimsy, and sheer size. But with a prominent and central location, standing apart but not isolated from the other grand buildings that don't quite surround it, but rather are designed around it, that needle marks the spot where the Academy begins. A humble structure, but tall and strong, carved out of granite, named for a man whose name is synonymous with courage to those who know his story, the monument to William Lewis Herndon represents the values of the mission of the United States Navy.

But what gives this particular monument its special, iconic status?

One day, a second-year midshipman looked at it and said, "I have an idea! Let's make the plebes climb it and put a cover on top!"

"That's a great idea!" an overeager classmate exclaimed. "But first, let's cover it with grease!"

Of course, it wasn't that simple. The annual ritual of the Herndon Climb, which has changed little in recent decades, arises from a progression of various earlier customs, all involving newly liberated plebes celebrating the end of the academic year and all centering around the monument, contributing to its special place in Annapolis lore and its own unique history.

In this book, we will try to tell the story of Herndon, shared by the men and women, midshipmen, and their families whose lives were touched by Herndon, the man, the monument, and the Climb.

As a spectator sport, the annual Herndon Climb offers all the excitement of a clash on the gridiron or in the boxing ring, with the added benefit of seeing, at the end of the contest, that everyone emerges a winner. The participants themselves, those young men and women battling the final challenge at the end of a year full of them, experience nothing less than a life-changing event. What may happen throughout those lives, no one can say, and only time will tell; but one thing they will have in common is that the Climb is a pivot point in all of them.

If you have never watched a Herndon Climb in its entirety, we highly recommend the experience, and we expect that this book will convince you to seek it out. Video recordings of modern Climbs are readily available, but nothing beats attending the event in person. You don't even need a ticket! But, be forewarned, early birds get the best viewing spaces, and they tend to arrive at first muster. But if you can't make the trip to Annapolis during Commissioning Week, a live video feed is available online, and it's the next-best thing to being there. If you're not a former midshipman but you've seen the Climb or have some familiarity with it, reading this book will deepen your understanding of the profound significance of the ritual and show you just how widely held are the shared memories of overcoming that seemingly impossible challenge.

If you have experienced a Climb for yourself, along with your plebe class, recently or decades ago, we hope this book will help you relive some of the excitement of it, without all the sweat and grease and people standing on your shoulders.

Jim McNeal
Hanover, Maryland

Scott Tomasheski
Los Angeles, California

1

THE HEIGHT OF GLORY, PART 1

(or, The Light at the End of the Tunnel)

Me mother was a mermaid, me father was King Neptune.
I was born on the crest of a wave and rocked in the cradle of the deep.
Seaweed and barnacles are me clothes.
Every tooth in me head is a marlinspike; the hair on me head is hemp.
Every bone in me body is a spar, and when I spits, I spits tar!
I'se hard, I is, I am, I are!

—Nonsensical verse, traditionally memorized by all first-year midshipmen at the United States Naval Academy. From *Reef Points: The Annual Handbook of the Brigade of Midshipmen*

Just imagine, for a moment:

You are eighteen, nineteen, maybe twenty years old. You have reached the end of the long, endlessly challenging academic year, your first at the United States Naval Academy. Final exam grades have been posted, and since you are accustomed to getting nearly straight As at the high-school level, your first-year scores in Academy classwork are nothing to write home about.

That is, if they had been achieved at any institution in the country other than the crucible of intense competition, exacting standards, and rigorous physical training at the Naval Academy, where the classrooms demand as much dedication to the study of chemistry and calculus as the field exercises do to the mastery of seamanship and hand-to-hand combat.

Now you are standing with your classmates, among the other freshmen, or "plebes," from your own company. They have been your roommates, teammates, shipmates and friends, brothers and sisters, since Induction Day. Together, you stand ready to take on the final challenge of the year. Today, you and your classmates will perform one of the Academy's longest-standing rituals: the Herndon Climb.

On your first day of your very first class in the required core program for first-year midshipmen, an English Literature and Rhetoric class, taught by an ancient civilian professor, you sat and wondered how in the world any of this stuff could possibly apply to a Navy career, when you suddenly and accidentally learned something, and it's stayed with you all year long. Just like when your high school English teacher said if you're joining the Navy, you should read Melville's *Moby-Dick*; you'll learn more about ships and the sea than you will in a hundred classes at the Academy.

She was probably right. But on that first day in English class at the Academy, you learned the meaning of the word "plebe."

It's short for "plebeian," a term that has its origin in ancient Roman times. A plebeian was not a slave, but rather a low-status, unprivileged citizen compared to the patricians, the ruling-class families, the elite and powerful. Next to a patrician, a plebeian wasn't exactly subservient, but by their dress and behavior there would be no doubt which one held the higher status. A Roman citizen was either a plebeian or a patrician, to be determined and assigned in strict accordance with his or her family's status, and that status would never change from birth until death. Whether a person was of the ruling class or the lower class had nothing to do with personal qualities or achievements, or even necessarily the accumulation or loss of wealth and property. Your status was nothing but an accident of birth.

That's funny, you thought. Why don't they go all the way and just refer to you and your classmates as slaves? At least the plebes in ancient Rome had *some* rights, *some* personal freedoms. Here you have been subjected to the random whims and inconsistently enforced discipline of the Academy's own version of arrogant, smirking, rule-crazy patricians: all the upperclassmen in the entire brigade.

You noticed one other, key difference between the Roman notions of plebes and patricians and those embedded in modern military academy customs. Here your status as a plebe was not an accident of birth, it was a purposeful and deliberate placement in the lowest status class, from which you were expected to graduate and achieve the vaunted high status of an upperclass midshipman. All you had to do was complete an extraordinarily challenging program of physical and academic training.

Then, you and your classmates had one more task to perform, the ultimate group exercise, before being *officially* pronounced "Plebes no more."

You will be allowed on the hallowed ground that has been previously untouchable for all plebes as you approach the twenty-one-foot-tall obelisk of smooth granite that stands in the Yard, just off Blake Road, where you've walked thousands of times in the last ten months. An hour ago, you formed company ranks in Tecumseh Court, organized by your score in the Sea Trials, and then took your places as companies along the path that leads from Tecumseh to Herndon.

Herndon.

You can't recall actually having seen the name on it, unless you saw it as a tourist, when you were a kid and you visited Annapolis with your family. *Reef Points* says it is called the Herndon Monument, and also says that midshipmen call it simply "Herndon." It stands at the head of a maze of neatly landscaped, monument-filled, diagonally traversing pathways on which you have heretofore been forbidden to walk, simply because they are diagonally traversing and plebes may not walk on any such pathway. From Blair Road in front of the Chapel, you have seen "September 12, 1857" on the visible side of the monument, but on the far side, you've been told, are simple raised letters spelling that single word.

HERNDON.

Today, you will not only see the far side of the monument for the first time. You will actually be able to *touch* the smooth granite.

And not only that, you will be able—in fact, you will be expected—to get a single one of your classmates to the top of the monument. Maybe it will be you. But there, at the obelisk's apex, twenty-one feet from the ground below, that classmate, or maybe you, will find a Dixie Cup hat, the simple sailor's cap, a taunting reminder of your low, plebe status.

The task is sometimes called the Herndon Climb in newspapers and such, but just as is the case with monument itself, the Climb is most often referred to simply as "Herndon."

To complete the task, the Dixie Cup must be removed. And then, in a gesture of simple dexterity but grand significance, the proper hat of a Navy officer, a peaked cap generally referred to as a combination cover, is used to crown the monument in its place.

Then, and only then, your status at the Naval Academy changes. The lowly plebe rank will be discarded, and you will be a fourth-class midshipman, with all accompanying rights and privileges fully granted.

So a certain amount of anticipation to undertake the task, complete the challenge, and get a combo cover to the top of Herndon has afflicted you and the other plebes, as you stand for what seems like hours while the superintendent prattles away on the steps of the bandstand. He is using a microphone and a suitable public address system, but you couldn't make out more than a word or two of the distant mumblings if you tried. And you aren't trying. You just want to *go*.

Ahead of you on the path, closer to Herndon than to Tecumseh, at the very head of the line, the plebes of the Iron Company, winners of the Sea Trials, are nearest to the action. But they can't hear the supe either. A few of them are already holding combo covers, spinning them on their fingers, flipping them around, tossing and catching them like Frisbees. In cocky anticipation of one of their own company being the first to scale the monument today, they have written their victorious company number across the tops of those covers, proudly but crudely, in heavy lines of Sharpie ink.

You have been ordered to remove your sneakers, which bear the telltale signs of heavy wear, if not abuse, and deposit them in a pile along with your classmates' similarly maltreated footwear. You've been told that you can collect them again after the conclusion of the Climb, or they will be donated to a local homeless shelter or other charity; you wonder who would ever want such a tattered and odiferous gift, but as a symbol of the trials of plebe year, you are willing to bid them farewell, and good riddance. You keep your rough, heavy, Navy-issued socks, for the moment, understanding that they might be useful in the impending effort to scale the monument; perhaps they will be balled up and thrown as projectiles, hopefully to dislodge the Dixie Cup hat from its taunting peak. You are allowed to carry nothing else on your person or in your pockets, no tools or harnesses or climbing gear, no ropes or nets or gloves. So, as you stand waiting, vibrating in anxious anticipation, many of your male classmates have already removed their T-shirts or torn them to shreds, and the women are hardly more modest, in blue swimsuits to accompany their identical shorts; this is the state to which the official "uniform" of the Climb has been reduced. Suntanned skin is an extremely rare sight among this group, but rock-hard muscles are not, and you realize in a moment of pride that you and your classmates are some of the best-conditioned athletes, or soldiers, or sailors, on the planet. This has not escaped the notice of the spectators, particularly family members and other civilians among them.

A low, temporary fence separates the plebes from the throng of spectators who have turned out to witness the Herndon Climb. As you stand on the brick pathway, waiting eagerly, you notice a few familiar faces among the crowd.

They are upperclassmen from your own company. And a few of them have somehow taken a special interest in you; they have been your tormentors, seemingly devoted to the singular task of breaking down your spirit. *Your* spirit. How you incurred such wrath you have no idea. You can't recall specifically, because it is now such a blur of memories, but you remember your first day at the Academy and the swearing-in ceremony; directly after the ceremony you would make your way to your room in the massive labyrinth of Bancroft Hall, where you assumed you would have a chance to rest for a little while. Right?

Wrong.

No sooner had you walked out of Tecumseh Court as a newly sworn-in midshipman and directed your eager steps to Bancroft Hall, than you were waylaid. It was your first confrontation with one of your major nemeses. She would become well known by the plebes in your company—perhaps *too* well known—by two entirely different nicknames. The women would call her Mary Madface, while the guys would refer to her as Betsy B. Buster, and opinion was sharply divided, along strict gender lines, as to whether the taskmaster Mary/Betsy reserved her most special venom and genuinely malicious cruelty for male or female plebes.

First, she demanded answers to specific questions, such as "what's the menu for tonight's meal?" and "what are the six rules of dead reckoning?" and "on what date did the *Bonhomme Richard* engage the British fleet?" and other queries on matters ranging from the trivial to the obscure.

If you are a serious history buff, dedicated to memorizing certain important dates in United States history, there is a slight possibility that you knew that the *Bonhomme Richard*, under the command of John Paul Jones, met the British on September 23, 1779. But you cannot have known the correct responses to the other questions; and in your first taste of military justice and Academy-style discipline, which often smacks of unfairness, verbal punishment is meted out for your lack of knowledge. Your salute is hypercriticized, your uniform somehow falls short of her expectations, your military bearing is dismissed as wholly inadequate. Finally, mercifully, she dismissed you, and just when you thought you were free, you met Harry Hairchest. Then Jimmy Jarhead. And others. Each one in turn amazed you with his or her thoroughly encyclopedic expertise in being a king-sized pain in the neck, and a few other body parts, as well. Eventually your upper arms and shoulders felt like jelly after performing multiple sets of pushups and holding salutes for what seemed like hours.

You and each of your roommates would become acquainted with a specific tormentor. You're in fine company, and your experience is common enough. Famous and distinguished Academy grads like Roger Staubach, class of 1965, legendary NFL quarterback; and John McCain, class of 1958, presidential candidate and six-term United States senator, both wrote in their autobiographies about their own upper-class tormentors during plebe year. Eventually, as the long year wore on, you came to believe that Harry or Jimmy or Mary/Betsy exceeded the boundaries of decency, going far overboard in enforcing discipline and pressing the limits of your military fortitude. You understood that your strength of character and self-control would be tested, but Harry and the others simply went too far, almost gleefully cruel and inhumane in their methods.

Funny, but both Staubach and McCain said the same thing.

And now, you know that this is a *very* special day indeed. For the first time, as far as you have ever seen, Mary Madface is *smiling*. And Harry Hairchest is clapping his hands. You have seen him do this before, but it was always in more of a taunting, insulting fashion, as he stood over you while you were engaged in some physically demanding task, like crawling under a barbed wire fence, mounted horizontally just inches above a stinking mudpit in one of a seemingly endless series of obstacle courses, while informing you pointedly that his ninety-year-old grandmother moves faster than you. Now, his clapping has more of a quality of—dare you say it?—something approaching encouragement. Maybe, even, a modicum of congratulatory approval.

Suddenly the report of a cannon blast, a blank shot from the small, relatively portable gun that is designed for ceremonial purposes and maximum noise rather than actual firepower, sounds from across the campus. Then a great, whooping cheer arises from your classmates and from the mass of spectators who have gathered to witness this event, your final struggle and ultimate triumph. Apparently, the superintendent has finally stopped yammering and has given the signal for the Climb to begin, evidenced by the startlingly loud boom of the cannon, and now, you and the other plebes finally rush forward.

There is a bit of pushing and shoving, all good-natured of course, as you move along the path. Soon there is open space in front of you, and as you get nearer to Herndon, the human traffic has thinned out and your short, quick steps turn into a near full-out sprint. In the chaos of the rush, one of your classmates isn't quite watching where she's going, and she runs smack into a light pole. You stop for a second; she seems to have been knocked nearly cold, but there isn't much you can do for her. Her close friends are helping her to a dazed, sitting position and summoning the on-scene medics. She will be in good hands, but it seems unlikely that she will be the one to reach the top of the monument today.

Finally, the twenty-one-foot-tall obelisk is in your sights. For the next little while—maybe just a few minutes, maybe an hour, or two, or longer—you and your plebe mates have a new nemesis, replacing Harry and Mary and the other upperclassmen.

That nemesis is called Herndon.

(To be continued . . .)

2

COMMANDER HERNDON'S MONUMENT

(or, How Herndon Earned It)

Commander William Lewis Herndon

1818–1857

Naval Officer—Explorer—Merchant Captain

In command of the Central America, home-bound with California gold seekers, Commander Herndon lost his life in a gallant effort to save ship and lives during a cyclone off Hatteras, September 12, 1857.

"Forgetful of self, in his death he added a new glory to the annals of the sea."—Maury

—Inscription on the bronze plaque at the base of the Herndon Monument

GREAT AND LONG-STANDING TRADITIONS typically have tangled, obscure origins and convoluted histories of development. The annual Herndon Climb at the United States Naval Academy is no exception.

One might argue that the origins of the Climb can be traced to the time of the ancient Egyptians, who erected the world's first stone obelisks, popularizing the form as a significant advancement in architecture and artistic design. The obelisk threads its way through history to modern times, where we find the Washington Monument and the Herndon Monument, among others, as recent examples of structures that are largely indistinguishable from the famous "Cleopatra's Needle," or needles, which were quarried in ancient times and today stand in New York's Central Park, London's Victoria Embankment, and the Place de la Concorde in Paris; or the massive "unfinished" obelisk that lies in situ in the granite quarry near Aswan, Egypt, and remains a major tourist attraction and source of fascination to anthropologists and archaeologists.

Customs that acknowledge the successful conclusion of a period of training and indoctrination of newly recruited soldiers or warriors and that link such a period to the practice of specific ceremonies and trials, date to earlier times. Military historians, however, can only speculate, rather than demonstrate, that such rituals were common in the Neolithic period among the first hominids who grasped stone hand axes and went into battle against rival tribes.

A convenient starting point for a discussion of the origins of the Herndon Climb lies roughly 2,500 miles to the west of Annapolis. There, on the morning of January 24, in the year 1848, near the modern-day town of Coloma, California, a man named John Marshall espied a peculiar, shiny lump of metal under six inches of water in the millrace of an unprepossessing sawmill on the south fork of the American River.

Of course, the lump of metal turned out to be gold, and there was much, much more of that precious metal to be found in the nearby hills, rivers, and streams. Despite the efforts of Marshall and his friend John Sutter, who owned the sawmill and large tracts of the surrounding land, keeping the discovery a secret turned out to be an impossibility. Word of Marshall's incredible find spread quickly to San Francisco, to the eastern United States, and around the world.

Soon the California Gold Rush began in earnest.

At the time, California had not yet been granted statehood in the Union; in January of 1848 it was not even a United States territory, but rather a part of Mexico. Its territorial status would soon be decided, however, as the Mexican-American War had recently been concluded, and the Treaty of Guadalupe Hidalgo would ultimately cede possession of California to the United States, along with all or part of nine other eventual states of the Union. In a bit of historic timing that seems almost providential, the treaty would be signed just nine days after John Marshall's discovery of that gleaming nugget at Sutter's sawmill.

Dreams of quick and easy wealth drove hundreds of thousands of individuals to change their career plans. Scores of subsistence farmers, overworked laborers, struggling merchants, and the like abandoned their former lives and lifestyles, and in many cases their families, to embark on an oft-perilous quest and a hazardous journey to the goldfields of California. Many carried little more than the clothes on their backs and dreams of new careers as prospectors and gold seekers. This floodtide of humanity was accompanied by an incalculable number of entrepreneurs who sought to capitalize on the frenzy by providing goods and services, including quack medical care, dubious legal practices, and prostitution, to the throngs of immigrants arriving daily from ships in San Francisco Bay, or by a variety of poorly established overland routes from the East Coast.

For the vast majority of the immigrants, hardships and disappointment were all they found. Stunned by the hyperinflation of prices for simple commodities like flour and oil and unprepared for the difficulties of life in the mining camps alongside stream beds and unsheltered rocky outcroppings that yielded precious little but discouragement and the occasional, tiny, mocking nugget of iron pyrite—"fool's gold"—or imagined glitterings that were nothing more than tricks of the light and the overhopeful mind, they eventually returned to their faraway homes with no gain on their heavy investments in human and financial capital.

Some fared even worse, losing everything, including their lives.

A precious few of the immigrants achieved their golden dreams, typically through a combination of backbreaking labor, a fair degree of engineering skill, entrepreneurial savvy, and a heavy measure of the most important factor of all: plain, blind luck. There was gold in the hills, for sure, but the deepest and richest of its veins were difficult to tap. Many of those with the resources to wrest it from the ground did so, and their endeavors were rewarded with enormous fortunes. The city of San Francisco grew from a sleepy seaport into a haven of the rich and famous, enriched by the new wealth. And gold came out of California, carried and jealously guarded by wealthy travelers who needed passage on ships, bound for their home towns, via ports at New York and other East Coast stopovers.

The Pacific Mail Fleet was the first of two government-subsidized steamship lines designed to address the exploding demand for increased passenger service and transportation of goods. Ships operated continuously, plying both coastlines, connecting New York to the Isthmus of Panama on the Atlantic side, and Panama to San Francisco on the Pacific. Travelers initially used a variety of means of conveyance, such as donkeys and canoes, to cross the relatively narrow isthmus, transferring to steamships on the opposite coast. Later, a fairly reliable train service became available, increasing the safety and comfort of what was once a far less pleasant journey traversing the Panamanian jungles.

By 1850, the rush for gold was in full force, and passage on board ships heading from New York to Panama and from Panama to San Francisco was in much higher demand than for that heading in the opposite direction. San Francisco Bay became clogged with abandoned, derelict ships as the lure of golden riches just a few miles away and rumors of gold nuggets lying openly in plain sight, there for the taking, proved too tempting for typical seamen; entire crews slipped away from their posts and headed into the hills. Even the ships of the U.S. Navy were not immune to desertion and abandonment; despite public floggings and a handful of executions of captured deserters, the Navy eventually threw up its hands and abandoned all hope of establishing a permanent presence in San Francisco until the Gold Rush had slowed and gold fever began to cool.

The westward trend of human traffic would begin to reverse itself within a few years, as throngs of the defeated and dejected headed for home, spending the last of their funds to book passage with the barest accommodations. Among them were the lucky few who found their fortunes in gold, and they paid for the more comfortably appointed first-class cabins. They carried their gold in many forms, including newly minted bars and coins, on board Pacific Mail steamships like the SS *Central America*.

The *Central America* was a sidewheel steamer, 280 feet long, weighing almost two thousand tons, and a veteran of service on the Atlantic, having made scores of transits between Panama and New York, typically with one stopover in Havana on each. Sidewheel steamers were the workhorses of the merchant marine: sturdy and spacious, built to carry passengers and cargo along major rivers and coastline routes, their contributions to the growth of the United States and the pre–Civil War economy are incalculable. With two huge piston-driven paddlewheels powered by enormous coal-fired boilers, the sidewheel steamer represented the state of the shipbuilding arts, incorporating significant technological advancements over earlier cargo ship designs.

On the morning of September 3, 1857, the *Central America* shoved off from the port of Colón, near Panama City, bound for Havana, and from there onward to New York. On board were nearly six hundred people, including passengers and crew, and mountains of cargo, including nine tons of California gold in carefully weighed bags of dust and raw nuggets, neat stacks of beautifully minted coins, and pallets of heavy golden bars and bricks, catalogued and precisely number-stamped at the burgeoning new San Francisco Mint.

And at the helm of the *Central America* was William Lewis Herndon.

He held the rank of commander in the United States Navy. He had twenty-nine years of experience in his sea chest when he took command of the *Central America*, his appointment being a consequence of federal legislation that required merchant vessels to be captained by Navy officers; therefore "Captain" Herndon refers to his specific position as senior officer on board a seagoing vessel, and "commander" is his actual Navy rank. Both titles are used interchangeably, and acceptably, in lore and documented fact.

Commander Herndon, captain of the SS *Central America*, did not have the invaluable advantage of modern-day weather forecasting information when he gave the command to weigh anchor in Havana and embark for New York harbor. If he knew that a Category 2 hurricane was brewing off the Carolina coast, he certainly would not have sent the *Central America* on a course to intercept it. Instead, he would have remained safely anchored in Havana, tamping down any uprisings among impatient crew and passengers eager for home ports, risking mutiny against the certain path of destruction that awaited them if they embarked on their charted course.

The passengers and crew on board the *Central America*, including her captain, understood something about the hazards of sea travel in those days. Everyone seemed to know someone who had perished in a shipwreck or had been separated from one by a minimum of degrees. Herndon and his crew were accustomed to hearing talk of storms and wrecks among passengers, and typically they did their best to discourage such worrisome thinking. But on September 8, as they embarked on what would be the final voyage of a great and noble vessel, placing their own lives in dire peril, speculation about hurricanes and ship sinkings took on an oddly prescient quality.

The *Central America* headed north, steaming through the Caribbean and skirting the coast of Florida, on a collision course with an Atlantic hurricane that manifested the terrifying powers of Neptune on a bad day; sheets of pounding rain, sail-shredding winds, and towering seas awaited Herndon and his crew. On September 9, the fateful appointment was kept, and the skies darkened and filled with buckets of rain as the first wave of a punishing storm assaulted the *Central America*, vulnerable and alone on an open ocean, eighty miles from the nearest safe harbor on the Carolina coast.

They tried to outrun it, outlast it, and outmaneuver it, but the hurricane hung over the *Central America* and battered the ship for three terrible days and nights. Colossal waves crashed on her open decks, flooding her causeways and sending floods of seawater into the ship's boilers. Maintaining the coal-fired furnaces of those boilers was of paramount concern: if they lost the fires of the boilers, they lost propulsion, and the two huge paddlewheels that drove the ship would be useless. If that happened, the *Central America* would be at the mercy of relentlessly pounding seas, unable to maintain any heading into the onrushing waves, which was her best chance of survival, instead of helplessly turning her flanks to the tempest, surrendering, releasing her claims on the world above, and slipping into a watery grave.

Bucket brigades ran tirelessly, day and night, passengers and crew alike pushing themselves beyond the limits of human endurance. The buckets ran along a line of men, and a few women, that snaked upward bearing loads of seawater, and downward with anthracite coal to feed the insatiable fires of the *Central America*'s massive boilers.

Eventually, the terrible forces of nature and weather overcame the ship and the utmost efforts of her passengers, crew, and her captain. Of the 578 lives on board the *Central America*, fewer than a hundred were saved. But Herndon was not among them.

William Lewis Herndon gave everything, including his life, in an honorable bargain, working until the very end, with every iota of his considerable skill and experience brought to bear to transfer as many of the *Central America*'s crew and passenger complement as possible to safety on board other vessels that came to assist, their own captains placing those vessels at great risk. Herndon wasn't the first, nor would he be the last, to fulfill a noble tradition of the sea: if his ship sank beneath him, the captain would be the last man on board.

Herndon was born in Fredericksburg, Virginia, on October 25, 1813. He would spend much of his life in the Navy and at sea. He married Frances Elizabeth Hansborough, and together they had a daughter, Ellen, who would go on to marry the future President of the United States, Chester A. Arthur. Herndon lived a fascinating life of adventure and sea voyages, commanding the Navy brig *Iris* in the Mexican-American War. When he died at the helm of the foundering *Central America*, Herndon was just forty-four years old.

"Forgetful of self, in his death he added a new glory to the annals of the sea." These words are attributed to "Maury," and they can be found at the base of Herndon's monument, inscribed on a heavy brass plaque. Matthew Fontaine Maury was Herndon's cousin, brother-in-law, and friend, and he was a key player in the effort to memorialize Herndon's sacrifice by placing an enormous granite obelisk dedicated to his memory on the grounds of the recently established United States Naval Academy at Annapolis, Maryland.

Maury undertook the task of forming a Herndon Monument Committee, raising the necessary funds to contract the Quincy Granite Railway Company of Quincy, Massachusetts, to deliver a piece of granite, designed and engraved according to a set of precise specifications. The terms of the contract were met on June 14, 1860, when the delivery arrived at the Academy, and the committee paid the contracted sum of 1,161 United States dollars. The monument to William Lewis Herndon was dedicated on June 14, and it has never been moved from the spot where it was raised.

Neither Maury nor any of his associates who designed, funded, and erected that monument had any idea of the beloved tradition that would arise among the Brigade of Midshipmen, nor that one day each year, during Commissioning Week or shortly before, it would become the center of attention and would symbolize the struggles of each and every freshman class at the Academy.

"Explorer."

It is one of the three titles bestowed upon William Lewis Herndon, according to no less an authority than the monument itself. That this title receives coequal billing alongside "Naval Officer" and "Merchant Captain" and bears testimony to that phase of his career that originally brought him the most notoriety.

Herndon was ordered to perform a task that would have made the average officer shrink from his duty. "Explore uncharted jungles and rivers and tributaries. Keep a careful account of all your observations, including the results of all soundings, geographic surveys, and encounters with native and homesteading individuals. Take a few pistols, plenty of ammunition, a handful of carefully picked men, and large quantities of tradeable goods, gold, and currency. Survive on a spartan diet of available and indigenous ingredients. Travel the entire known length of the Amazon River and prepare a detailed survey and census upon completion of the journey."

The exact text of Herndon's written orders, direct from the Secretary of the Navy, differed from such a blunt and realistic description of the assignment. Nevertheless, then-Lieutenant Herndon departed Lima, Peru, at the Pacific terminus of the Amazon, on May 20, 1851, embarking on the adventure of a lifetime. For nearly an entire year, he and his second, Lt. Lardner Gibbon, and an ever-changing cast of locally recruited talent experienced the full scope of daily life in the Amazon Valley.

Herndon meticulously recorded his observations, per his directive from the Navy. He and Gibbon and their hired hands encountered an extraordinary variety of native flora and fauna, including everything from carnivorous plants to vampire bats, and a menagerie of exotic species largely unknown in Europe and North America. They captured live specimens and skinned and dissected others, and they prepared extensive reports and diagrams that added immeasurable value to the scientific understanding of natural history and diversity.

They took depth soundings and measurements, charting the navigability of the local waterways, including the main branch of the Amazon and its hundreds of tributaries. They scaled mountains, some as tall as sixteen thousand feet, measuring those altitudes by the precise boiling point of water. They came into extensive contact with villagers and farmers, indigenous people and European expatriates; they toured gold mines, silver mines, copper mines, and rubber-collecting operations. Per his orders, Herndon kept a careful accounting of the size and economic productivity of all these endeavors, surveying entire towns and villages and recording their population sizes and the current states of the local economies.

A man of his time, a firm devotee of the principle of "manifest destiny" and mankind's directive to bend the forces of nature to its will, Herndon bemoaned what he perceived as untapped economic potential across vast, sparsely populated regions rich in mineral wealth, timber, and plants with medicinal and food value. He found many of the residents of the Amazon Valley to be afflicted with laziness, sloth, and drunkenness, which he attributed largely to the warm and humid climate. "Men in these countries are not ambitious," Herndon eventually wrote in his report. "They are not annoyed, as the more masculine people of colder climates are, to see their neighbors going ahead of them. They are contented to live, and to enjoy, without labor, the fruits which the earth spontaneously offers. I have very little doubt but that the bold and ambitious Frenchman, and the hardy, persevering, scheming American, who likes little that anyone should go ahead of him, would alike, in the course of time, yield to the relaxing influence of a climate that forbids him to labor."

His further observations are startling to the modern reader. "Let us now suppose the banks of these streams settled by an active and industrious population, desirous to exchange the rich products of their lands for the commodities and luxuries of foreign countries," he wrote. "Let us suppose introduced into such a country the railroad and the steamboat, the plough, the axe, and the hoe; let us suppose the land divided into large estates, and cultivated by slave labor, so as to produce all that they are capable of producing. And with these considerations, we shall have no difficulty in coming to the conclusion that no territory on the face of the globe is so favorably situated, and that, if trade there is once awakened, the power, the wealth, and the grandeur of ancient Babylon and modern London must yield to that of the depots of this trade that shall be established at the mouths of the Oronoco, the Amazon, and the La Plata [rivers]."

And there's more: "Civilization must advance, though it tread on the neck of the savage, or even trample him out of existence. Throw open the country to colonization, inducing people to come by privileges and grants of land. I am satisfied that in this way, if the Indian be not improved, he will at least be cast out, and that this glorious country may be made to do what it is not doing now—that is, contribute its fair proportion to the maintenance of the human race."

These are hardly the sentiments of a modern, progressive mind, but they were quite common in pre–Civil War America and on the western frontier of the growing nation.

Herndon seems to have steered clear of such philosophizing as he navigated the Amazon Valley, where travel was by a variety of imperfect means: rickety canoes and small boats, mule train, and on foot. The nearly year-long journey was hardly an epicurean gourmet pleasure cruise; frequently, the daily bill of fare consisted of little more than boiled cassava tubers, known as *yucas*, and cold monkey soup; admirers and protectors of the lower primate species will be horrified to learn that "monkey soup" is exactly what its name implies.

Herndon and his men were constantly beset by food shortages, finding that local farmers and villagers were often loathe to trade with the party, claiming they had no livestock or produce, despite the obvious presence of chickens, pigs, cattle, and abundant crops, stores of ripe sweet plantains, and nourishing *yucas*, corn, and beans. Fearing bandits and robbers, the locals frequently claimed they had nothing to sell or trade, hence nothing to steal or plunder.

As a further vexation, all along their route Herndon and his company encountered men who seemed in need of manual employment, as porters, guides, cooks, and the like, but time and again the travelers found themselves plagued by a scarcity of men willing to work for wages. To an exasperated Herndon, it seemed that much of the population of the valley maintained a level of alcoholic intoxication, rendering them largely useless, each and every day from sunrise to sunset and beyond, as they preferred jars of fermented *yucas*—another use for the versatile tuber—to any form of labor.

Despite the privations, difficulties, and unanticipated shortages of food and ready labor, the expedition soldiered along. Early on, Herndon made the critical decision to separate from Lieutenant Gibbon, dividing their huge task and essentially doubling the amount of data that ultimately informs their report. Under the new plan, Herndon and Gibbon split their resources, supplies, and funds, with Herndon to explore the original planned route along the headwaters and main trunk of the Amazon, and Gibbon tasked with exploring the Bolivian tributaries.

"It was a bold, almost rash determination, for the party seemed small enough as it was; and we might readily encounter difficulties on our route which would require our united exertions to overcome," Herndon wrote. "I had many misgivings, and I told Gibbon at first that it seemed midsummer madness; but the prospect of covering such an extent of territory, of being enabled to give an account of countries and rivers so little known, and the reflection that I need not abandon routes that I had looked upon with a longing eye were so tempting that they overrode all objections; and we set about making preparations for the separation." They would reunite in Pará, Brazil, and combine their efforts.

Gibbon was an outstanding officer in his own right, probably deserving of a monument of his own, as tall as Herndon's. Among Gibbon's more esoteric skills were those of a talented sketch artist, and the dozens of sketches he prepared during his travels add terrific interest for the modern reader of the final, published manuscript. Ultimately Gibbon would produce his own, independently authored version of his personal observations.

When the original manuscript finally reached the desk of Navy Secretary John P. Kennedy, the secretary might have been surprised enough to drop his cigar in his beer glass.

Exploration of the Valley of the Amazon, 1851–1852, under the proud byline of William Lewis Herndon, was hardly a dull and dry collection of farm inventories, mining-operation balance sheets, and actuarial tables. It was much, much more. Readers were captivated by the prolific prose that flowed from Herndon's pen, by his vivid accounts of exotic life forms, and especially by the surprisingly self-deprecating sense of humor that infuses the work. Gibbon's sketchbook of detailed drawings took it to another level of entertainment value; more than a century and a half later, it still reads like a thrilling adventure novel, in the style of a James Fenimore Cooper or Alexandre Dumas.

Soon, the few hundred copies that were printed for the secretary and other Navy brass, members of Congress, and a few other interested parties had been passed hand to hand with the "you gotta read this" enthusiasm that accompanies a modern underground blockbuster. An additional print run churned out another ten thousand copies, followed by a third run of thirty thousand, and the book became an international bestseller.

Among the book's many contemporary aficionados was a young Missourian named Samuel Langhorne Clemens, whose adventurous spirit was fired to the point where he set out to duplicate Herndon's journey. He left behind his mundane employment as a shopkeeper and embarked on the Mississippi River, destined for New Orleans, where he intended to book passage on the first vessel bound for the state of Pará, in Brazil, at the mouth of the Amazon. From there, he intended to undertake the journey upriver into the headwaters high in the Andes, in the opposite direction of the downriver course followed by Herndon. Arriving in New Orleans, Clemens was disappointed to find no vessels bound for Pará; nor were any scheduled, at any time, and he was reluctantly forced to abandon his dream quest. Nevertheless, he would eventually live a lifetime full of adventure, writing and publishing dozens of novels and nonfictional accounts under the pen name "Mark Twain." Most are well-known American classics. That such Twain works as *Innocents Abroad* and *Life on the Mississippi* are clearly influenced by Herndon is hardly a matter of debate among literary scholars.

Clemens never made it to Pará. He earned many honors and accolades, and he inspired many tributes, plaques, and paintings. His mustachioed likeness may be found on countless book covers, posters, and T-shirts, but no one ever erected a twenty-one-foot granite obelisk in his memory.

3

THE SNAKE DANCE

(or, How This Whole Crazy Thing Got Started)

Let the word go forth . . . the torch has been passed.

—John F. Kennedy, Inaugural Address

The Naval Academy Museum in Preble Hall is a storehouse of history, full of treasures great and small. A first-time visit to the Annapolis campus is incomplete without a viewing of the museum's collection of unique and priceless artifacts of singular historic significance. A presentation sword that was carried by John Paul Jones; the very table from the USS *Missouri* where the Japanese surrender of 1945 was signed; and the tattered but painstakingly preserved battle flag with the iconic words of Oliver Hazard Perry, "Don't Give Up the Ship," are beautifully presented alongside family-friendly interactive animated battle maps; extensive collections of antique weapons, uniforms, medals, artwork, and documents; and many other rare objects.

And that's just the *first* floor. The second level of the museum offers perhaps the world's finest collection of ship models, precious and wondrously detailed constructs of wood and bone and thread and cloth, full of tiny details worked with infinite patience by highly skilled fingers. It's a safe bet that any midshipman with a history buff for a parent or grandparent included the museum on that relative's first personally guided tour of the Academy.

Anyone with a detailed knowledge of the byways of Annapolis, such as schoolteachers who conduct regular class excursions and Academy docents and tour guides, knows the Naval Academy Museum. And anyone who knows the museum's history over the course of the last five decades knows Jim Cheevers. To call him a walking, talking encyclopedia of Navy history and Academy lore is to overrate the encyclopedia.

Jim became the museum's curator in 1967, and he continued in that post for the next fifty years, accumulating several ordinary lifetimes' worth of knowledge in the process, until finally stepping aside in 2017. Understandably, he takes a great deal of pride in the museum, and he is still a frequent, fatherly presence among the current staff. He can often be found examining the minutiae of certain favored objects for the umpteenth time, or scanning the archives for undiscovered golden nuggets of historical data.

"This is one of the oldest museums in the country. We have collection records here dating from the U.S. Naval Lyceum at the Brooklyn Navy Yard, founded in 1834, that was closed in 1888 and its collections transferred here," he says. "And the Boston Naval Library and Institute, 1842 to 1921, came here, too. So, we have a fabulous collection. The first major collection we got was the Navy's trophy flags." That priceless collection started with the flags of the captured warships *L'Insurgente* and *Le Berceau* during the undeclared naval war with France, otherwise known as the Quasi-War, between 1798 and 1800. Soon after that conflict, the United States Congress passed a law requiring preservation of these and all future trophy flags, eventually transferring them to Annapolis in 1849, before the Academy was officially christened.

Although he takes immense pride in all things Navy, Jim attended the College of William & Mary and served in the Army. With a degree in history, experience in museum work at Fort Benning, and a sergeant's rank, he was a strong candidate to replace the previous curator of the Naval Academy Museum, Capt. Dale

Mayberry of the class of '33, who stepped down after thirty years on the job.

Needless to say, there isn't much turnover at the curator's post, which also comes with a great deal of job satisfaction. Jim's only complaint, in fifty years, is in regard to the notorious shortage of parking spaces around the Academy campus.

"You can't help but love Annapolis, and the Academy was just a wonderful place to work," Jim reflects. "I met a lot of wonderful people over the years. Staff, faculty, midshipmen. In the beginning, it was lovely. Toward the end it got a little ugly—because of parking. I was getting up at five a.m., getting in here at least by six fifteen, just to get a parking place. My first five or ten years we had parking across Maryland Avenue for the museum . . . five staff spaces, of which I did have one. There was another time, under the Carter administration, where they made us pay for parking. Then we were promised, they told us, keep the records and we'll get reimbursed. We never did!" he says with a laugh.

Despite the parking woes, Jim Cheevers' five decades of curatorship brought immeasurable improvements and growth to the Academy Museum. He worked tirelessly to upgrade the museum's collections and displays, improve the visitor experience, and expand the depth and breadth of the archives.

One of his many additions to those archives is a bulging manila folder stuffed full of documents related to the Herndon Monument, and specifically to the Climb. In the midst of modern society's ever-growing reliance on electronic forms of communication and digitized storage of data, Jim meticulously combed through more than a half-century's worth of old newspaper articles, Navy newsletters, and the Academy's own *Lucky Bag* yearbooks, using an overworked Xerox machine to copy every relevant document he came across, building a file of historic importance and key significance for Herndon enthusiasts.

A long career of dedicated, passionate work by Jim Cheevers gave the Naval Academy one of the nation's great specialized museums, and his prevailing interest in the Academy's proud traditions prevented the history of the Herndon Climb from fading into obscurity.

Shipmates Forever is a nearly forgotten Hollywood motion picture that perfectly captures the United States Naval Academy in a cinematic time capsule between the two world wars.

A Warner Brothers film released in 1935, written by Delmer Daves and directed by Frank Borzage, its fictional storyline takes place in large measure at the Academy, and it boasts numerous sequences filmed at actual Annapolis locations. The film offers the viewer a generous look at such familiar sights as Bancroft Hall, Tecumseh Court, the Ring Dance, Monument Hall, the Chapel, Gate Three at Maryland Avenue, the waterfront, and many of the best-known monuments on campus, including Herndon. It is a typical 1930s Hollywood production, meant to entertain and cheer Depression-era audiences, so it has music, comedy, romance, and an exciting action climax cobbled together into a story about one young man's career at the Academy, from the trials and tribulations of plebe year, through his fateful, first training cruise as a senior.

Dick Powell, a major star of the 1930s and 1940s, known for both musical comedy and later tough-guy roles, plays Dick Melville, a young Broadway crooner whose father just happens to be a four-star admiral and superintendent of the Academy. Dick's girlfriend, played by Powell's frequent on-screen partner Ruby Keeler, is soured on the Navy in the wake of a series of family tragedies dating to World War I, and Dick has achieved moderate success as a nightclub singer and radio performer; nevertheless, he accepts his father's challenge and enters the Academy. He struggles through a typical plebe year with an upper-class tormentor who, knowing that Dick is a singer, orders him to sing in awkward situations in an effort to humiliate him. But Dick somehow always manages to come off well, even finding time to join his sweetheart for some typical Hollywood-style romantic high jinks, with a musical number or two thrown in for good measure.

The authenticity of the film shows clearly in scenes that portray the Tecumseh statue, which stands near Bancroft Hall. The iconic statue is a replica of the figurehead of the USS *Delaware*, portraying Chief Tamanend of the Delaware Indians, and over the years it has come to be known as "Tecumseh" and has inspired a number of popular superstitions, as seen in the film. At one point Dick stands in front of the statue, invoking its supposed wisdom as he struggles with a major decision about his future at the Academy; and later his roommates practice the long-standing tradition of throwing pennies at the figure of Tecumseh, this time a small replica statuette, hoping for luck and deliverance in the form of passing scores on their exams.

As an extra treat for Navy history enthusiasts, the film includes rare stock footage of the legendary dreadnought battleship USS *Arkansas*, launched in 1911 and a veteran of both world wars, as well as the famed Martin P3M-2 seaplane, of which only six were ever built.

The Herndon Monument appears in a key scene about halfway through the film. Dick makes it through plebe year, but he is cynical about many of the Academy's customs, not to mention its strict rules. Final exam grades are posted, the graduation ceremony has ended, and Dick's roommates and the other plebes are happily swarming around the recognizable granite obelisk of Herndon.

Dick is in his father's office, which has a window directly overlooking the monument, inexplicable except in a Hollywood movie.

"Why aren't you with your classmates?" his father asks, looking at the celebration below.

"Oh, I don't know," Dick replies. "Some of these traditions seem like horseplay to me. There's nothing in the book of regulations that says I have to dance around Herndon Monument just because I passed plebe year."

This leads to a dramatic confrontation that underscores the remainder of the film. With a bit of judicious searching, a determined film buff can locate this pivotal moment and find out exactly how the story plays out. But whether Dick will ultimately win his father's respect and his sweetheart's everlasting affection, and prove his mettle when his ship runs into trouble at sea on his first training cruise, are matters that are seldom in doubt.

The Herndon obelisk may be seen in only a few brief shots, but the historic antecedent to the modern Climb is on full display, recorded for posterity, as the plebes in their full, dark-colored daily uniforms rush and swarm around the monument, cheering and shouting delightedly, some of them standing up on the low pedestal at its base, triumphantly running their hands over the raised lettering that reads "September 12, 1857."

This wonderfully illustrative sequence from *Shipmates Forever* was undoubtedly staged for the movie cameras, since actors Ross Alexander, Eddie Acuff, and John Arledge, who play Dick's on-again-off-again friends and roommates, are clearly visible in the crowd of celebrating plebes. Nevertheless, the action portrayed on screen aligns closely with the following description of Herndon from the 1934–35 edition of *Reef Points*, describing the moment when plebes became "youngsters," or sophomores. "On graduation day this monument is the rallying point of the new youngsters, where, with many a cheer and cry of 'No more plebes,' they celebrate their new found freedom." Later in the volume, *Reef Points* continues: "A leg-straining time at graduation is followed by a mad snake dance about the Herndon Monument to complete the rituals of Youngsterization."

By 1935, the Snake Dance was already a decades-old Academy custom. At the conclusion of the annual graduation ceremony, when each class was symbolically elevated to the next highest level and the seniors became commissioned officers, the midshipmen who woke up in the morning as plebes were officially relieved of that lowly title. After maintaining the minimum passing grade average and avoiding the maximum allowable number of demerits, the newly liberated ones celebrated by rushing en masse to the Herndon Monument, treading for the first time on the grassy lawn surrounding it, a "no man's land" for plebes.

The term Snake Dance might conjure any of a number of images. Some may think of a conga line, the Cuban-inspired carnival dance and ubiquitous staple of modern wedding receptions that tends to break out spontaneously once partygoers have been enticed to shed a minimal level of inhibitions. Those who have visited the Indian subcontinent along tourist byways have likely seen a snake charmer, who plays a *pungi* or similar wind instrument in tones that seem to hypnotize the senses of a snake, typically a venomous cobra or viper, enticing the creature into a kind of swaying, rhythmic dance. Snake-charmer performances on streets and sidewalks are often accompanied by a human dancer, mimicking the snake's fluid motions along with the hypnotic musical lines. Ancient Near Eastern and Far Eastern cultures have traditional snake dances, typically involving a solitary, female, scantily clad dancer who incorporates sensuous, serpentine motions in the performance.

According to the unabridged version of the *Oxford Dictionary of the English Language*, a snake dance has multiple definitions. Anything from a Hopi tribal celebration that involves the handling of live rattlesnakes, to a high school Homecoming parade that circles a football stadium in a serpentine fashion can be a snake dance. The term was in its heyday early in the twentieth century, when a variety of college and high school celebrations, Homecomings, and pep rallies included some form of processional that was commonly referred to as a snake dance.

In the early years of the twentieth century, the Naval Academy developed its very specific Snake Dance custom, as seen in the film *Shipmates Forever.* The earliest evidence of the custom, when plebes celebrated their own symbolic promotion immediately after the completion of the graduation ceremony, appears in the 1910 edition of the *Lucky Bag* yearbook.

A single, grainy monochrome photograph illustrates the event: a long line of men in their dark uniforms, in a well-organized line that "snakes" past Zimmerman Bandstand and terminates in a great crowd around Herndon. The text, a first-person account of the history of the class of 1910, including their plebe year experience in 1907, reads: "June at last arrived and we sang 'God Be With You,' went through the graduation with ennui becoming to those who had already seen two, and finally swooped out to swarm around the Herndon monument, cheering everyone and everything, not forgetting ourselves."

The next edition of the venerable *Lucky Bag* offers this insight: "Impatiently we waited for the word 'dismissed,' and then what a race for the heretofore forbidden precincts of the Lane! How we sang and capered 'round the monument! The memory of that bright day is still with us, and perhaps none but our own graduation day can eclipse it." So reads the history of the class of 1911, recalling their triumphant moment three years earlier.

By then, the Herndon Monument was well-established as the focal point of the plebes' celebration. According to *Reef Points* for the 1910–11 academic year, "There are no traditions or customs for the Fourth Class to live up to, except, perhaps, to be 'seen and not heard.' They are officially barred from all hops, except the Farewell Ball. They have no desire to walk in Lover's Lane, this being an unnecessary pleasure. After graduation is over, however, they usually have a merry romp around the Herndon [M]onument, and, after that, no one cares whether or not they use the classic walk." This entry also appears nearly verbatim in the 1911–12 *Reef Points*.

Both *Reef Points* and the *Lucky Bag* yearbooks from the time period mention Lover's Lane, or Love Lane, or simply the Lane. These are fairly archaic terms, seldom heard nowadays, referring to the unofficial designation of the walkway on the Yard directly adjacent to Herndon, where midshipmen at the all-male Academy were permitted to meet female guests. Naturally, plebes were not permitted on this path or on the lawn around Herndon, and therefore the entire area became the scene of the Snake Dance parade, often called the "tain't no mo' plebes" ceremony.

The 1922–23 edition of *Reef Points* pays particularly close attention to the significance of Lover's Lane. "As one looks across the Yard from the Chapel, he sees Lover's Lane, a winding bit of gravel sacred to the upper classes since time immemorial," it reads. "On it are the Herndon Monument, calling to mind the brave Commander Herndon, who went down with his ship, the *Central America*, of the Panama Mail Steamship Line, on September 12, 1857, and the Japanese Bell, a gift to Commodore Perry in the far-off day when he opened Japan to the world [1857]."

The gravel was eventually replaced by a brick pathway, and the legacy of Commodore Perry's relationship with Japan would undergo paroxysmal changes over the ensuing decades that culminated in World War II.

By 1930, plebes reading *Reef Points* were advised as follows: "Lover's Lane. This bit of gravel walk, lying between Chauvenet Walk and Blake Road, and the benches lining it are sacred to upper classmen until the supreme moment immediately following graduation when the newly made 'passed Plebes' have all the right of way over all others to snake dance along the Walk and kick all the gravel off as their first act as youngsters."

An insight into the reason for the eventual demise of Lover's Lane as a meeting place for upper-class midshipmen and their romantic interests appears in *Reef Points* for the 1937–38 academic year: "Lover's Lane, the gravel walk that winds through the yard, is taboo to the Plebes until after graduation, when they hold their 'Tain't no more Plebes' snake dance and celebrate at the base of the Herndon Monument. Way back when . . . it was lined by a head high hedge that afforded a wonderful screen from peeping toms and scandal mongers." The 1941–42 edition put it more bluntly, stating that Lover's Lane "got its name somewhere in the dim, dark past when there was a high hedge on either side of it. The hedges are gone now, and Lover's Lane is just a trite name for a gravel walk."

By 1940, as the storm clouds of war threatened to draw the United States into an inevitable all-out conflict in Europe and the Pacific, the *Lucky Bag* yearbook shows the plebes of the class of 1943 actually forming a human pyramid and boosting up a classmate to perch at the very top of the Herndon Monument. By this time, a new sartorial wrinkle had become an integral part of the custom, according to *Reef Points*: "As they snake-dance down Buchanan Walk and up Youngster Cutoff with their coats turned inside-out so as not to show their lack of stripes they dance around this monument to signify to all the world that 'Tain't no mo' plebes.'" *Reef Points* of 1944–45 catches up with the relatively new idea of climbing Herndon: "Immediately after the graduation ceremonies on the last day of your plebe year, the Herndon Monument will be the scene of great noise and snake dancing. When the first man in your class reaches the top of this obelisk, located across from the Chapel, you will be 'No mo' Plebe.'"

So by the mid-1940s, the annual Snake Dance celebration was culminating in a physical climb of Herndon, with the end of plebe year and upgrading of status not pronounced official until a member of the class reached the top of the monument. Further refinements to the custom were added over the ensuing years as it became more similar to the familiar spectacle of today, with extensive media coverage and crowds of spectators.

Starting in 1947, the tradition of placing an officer's combination cover at the top of the monument began, representing the plebes' ultimate triumph over the adversity of their first year at the Academy. And 1962 was the first year that a Dixie Cup hat was added to the ritual, the exchange of caps at the apex further cementing the profound meaning of the Climb as an important step in the life of every midshipman.

Another seismic shift in the growing cultural significance of the Climb, to which midshipmen were increasingly referring simply as "Herndon," happened in 1966 and changed the entire landscape of June Week at the Academy.

The first few days of early June that were known for generations as June Week, culminating in the graduation ceremony for the senior class, would later be christened as Commissioning Week when the academic year was rearranged so that graduation would fall in May. And most importantly, the Herndon Climb, otherwise known as Herndon or by its official designation as the Plebe Recognition Ceremony, was for the first time scheduled to take place as one of the first events in the week, *before* graduation.

Now the Herndon Climb took on a new meaning, as plebes were released from the bonds of their low status with several days remaining before their senior class tormentors would be commissioned as officers in the Navy and leave their Academy entanglements behind them. And the youngsters, eager to don the shoulder boards of second-class midshipmen, similarly needed to wait until commencement exercises to take that privilege.

Those youngsters immediately understood the prominence of bragging rights regarding the Climb. And they would do everything in their power to make the task as difficult as possible for the plebes and deny them the taste of victory for as long as possible. So arose the practices of greasing the monument with enough lard to fry a mountain of dough and eventually resorting to the rule-bending practice of securing the critical Dixie Cup to the top of Herndon Monument with duct tape and industrial-strength adhesives.

With the death knell of the Snake Dance came the ascendancy of the Herndon Climb as a ritual in Academy culture, second only to graduation, when Presidents of the United States of America and secretaries of defense and chairmen of the Joint Chiefs of Staff are commonly featured as speakers.

And the Climb became an unforgettable experience for the generations of young men and women who experienced it firsthand.

4

A NEW WORLD RECORD

(or, The Great Chicken Wire Snafu)

> I wish to have no Connection with any Ship that does not sail fast for I intend to go in harm's way.
>
> —John Paul Jones

AFTER A FEW DECADES of evolution, by 1962 the most notable features that define the tradition of the modern-day Herndon Climb were firmly in place.

But still the ritual didn't carry the same weight of officially sanctioned ceremony that it does now, when the plebe class of '65 approached the monument for the first time, setting foot on the sacred grassy earth surrounding it and laying their hands triumphantly, if not reverently, on the cool, smooth granite. They had already been pronounced "Plebes no more" when final grades had been posted and the seniors had the chance to throw their hats in the air at their graduation ceremony.

When the Field House airspace filled with that mighty flock of combo covers, given flight by their proud class of '62 owners, impatient plebes immediately alighted and sped to their rooms in Bancroft Hall, to change from formal uniforms into daily work whites and commence the assault on Herndon.

In those days, scaling the monument and placing a cover at the top was more a matter of privilege than a requirement for advancement. After being denied the very right to touch the grass or tread upon the diagonal walkways in the vicinity of the obelisk that stood so defiantly in front of the Chapel, so temptingly within sight but out of reach every day of an excruciating plebe year, the freshmen were eager to undertake a final challenge, a symbolic gesture that represented their triumph and their implied promotion to third-class midshipmen.

But their upper-class tormentors, who had worked so hard all through the academic year to ensure that the plebes came by nothing without a demanding physical price, were disinclined to allow an easy path to the top of Herndon, and to that end, the custom of preparing the monument in advance of the Climb was well-established.

Detailed records are not kept, but over the decades a lengthy PX list of different substances, in varying amounts and combinations, have been applied to the monument just prior to the Climb, significantly increasing the difficulty level of the task. Lard, vegetable shortening, butter, olive oil, gun-metal lubricants, and experimental oddities from the Academy chemistry laboratories have all been used; in recent times, only the edible options have been selected, respecting the fact that whatever is used will end up covering the plebes' skin, getting in their eyes and ears, coating them from head to toe. Concerns about the potential toxicity of such petroleum-based compounds as cosmoline have led to their disuse as Herndon lubricants.

The sophomores tasked with preparing the monument for the class of '65 may have used cosmoline, a substance well-known to generations of soldiers and sailors as a rust inhibitor, commonly applied to the bare metal of moving gun parts, vehicles, and shipboard machinery. A heavy coating of this readily available material upon the smooth granite sides of the monument would certainly deter any notions of a quick and easy scamper to the top.

Then, one of those sophomores, or youngsters, came up with an idea. And it turned out to be one of the worst ideas in the history of the Climb.

As far as strategic military blunders go, it is comparable in folly, if not fame, to the Union Army's ill-fated deployment of the notorious Petersburg Mine, or to the fateful decision on the part of the commanders of ancient Troy, who defied all logical advice and brought an enormous Greek-built gift horse into their fair city.

The plebes of the class of 1965 were eager to pit themselves against the forces of nature and gravity and the monument. There was no cannon shot to begin the assault; instead, the moment of hat throwing by the graduating seniors served as permission for the plebes to commence the challenge, one that was an honor and a privilege; their very presence on any of the expanses of green lawn beside Blake Road meant that they had successfully completed an arduous first year at the Academy.

As the newly commissioned officers celebrated with their classmates and families, the plebes rushed to their rooms for a quick change of clothes and turned their attention to Herndon.

Decades removed, as Ed Linz examines photographic evidence of the event, his memories are as fresh as a dew-drenched flower on a Maryland spring morning.

Ed had been unlucky enough to be conscripted into duty as an usher at the graduation ceremony, and his room in Bancroft Hall was not ideally situated for a quick sprint from the Field House or from there to the Herndon Monument. When he arrived at the obelisk, in the shadow of Chapel Dome, the Climb had already begun in earnest.

"There was a group of my classmates that had already formed the bottom of a pyramid," Ed recalls, "and I jumped in and started climbing up a few levels."

Moments later, once Ed ascended those few levels, he found a countermeasure in place intended to hamper the plebes' progress: a great length of narrow-gauge chicken wire fencing, wrapped around the upper half of the monument. Beneath this tangled matrix, the granite surface was almost entirely free of grease, whereas the lower part of the monument was very heavily greased.

If the youngsters thought they had come up with a brilliant new strategy, one that would frustrate the plebes' attempt at a quick and clean Climb, forcing them to expend an inordinate amount of time and energy untangling and removing this knotty wire barricade, they were sadly mistaken; they did not anticipate the agility and abilities of Kentucky-born Ed Linz.

The typical suburbanite or city dweller might have been loath to attempt to climb a chicken wire fence, owing to the difficulty of comfortably gripping such narrow-gauge wire, but to Ed it was second nature.

"Growing up on a farm in Kentucky, I spent much of my life prior to the Academy climbing trees, extricating semidead squirrels that I had shot from the holes that they had gone into," Ed says with a chuckle. "I was very adept at climbing, and I was also fairly light. I probably weighed about 130 pounds or so at the time."

Ed wasted no time at all. He seized hold of the chicken wire, as he had done many times in his young life around poultry enclosures and rabbit hutches. But now, after a year of Academy training, Ed's body was carved to a rock-hardness, and his mind was tuned to a state of mental toughness that had him well-prepared to withstand a great deal of challenge and punishment. The inner surfaces of his hands and fingers may have been torn and bloodied by the fine steel mesh of the chicken wire, but this meant little in comparison to achieving the goal.

To the amazement of his classmates and the assembled spectators, Ed scrambled up the tangled mass of wire, and a moment later he was at the top of the monument, calling down for a combo cover, ready to place it at the apex and put this Climb into the history books.

"I called down for a hat," Ed recalls of the victorious moment. "Somehow, I was able to hold on with one hand, grab a hat out of the air, and lay that baby on there."

No one was more surprised than the sophomores, whose implementation of the chicken wire scheme had backfired catastrophically, leading to one of the very fastest Climbs in more than seven decades of record keeping. The total elapsed time from start to completion is recorded as exactly three minutes.

"I think there was a whole cadre of youngsters—third class, who were now second class—who were bitterly disappointed," laughs Ed, "because we were no longer afraid of those guys, we were making fun of them, and everything else that goes with it."

Some sources report that the chicken wire was actually a cargo net, thrown up by the plebes to aid them in their effort. Ed readily dismisses this notion and bristles at the idea that his class had somehow "cheated."

"Farm boys from Kentucky know chicken wire when they see it," he points out.

But the Climb wasn't finished with Ed, despite Ed's thinking he was finished with the Climb. Looking down after placing the combo cover at the apex of the monument, he saw the human pyramid fall away, and he found himself alone, clinging to chicken wire, his toes still a good fifteen feet from the ground. Worse, he realized that an unidentified classmate had been a counterweight on the opposite side of the monument, also holding himself on the wire fencing and keeping it balanced, relatively tight against the surface; and now that classmate had jumped, or fallen, to the ground, leaving Ed sky-high and in a precarious position as the fencing sagged and pulled away from the monument.

He doesn't recall if he had a hard fall to the grass below or if he was received by the cushioning arms and bodies of his classmates, most likely the former. And he was immediately buttonholed by a tenacious reporter from a local newspaper, eager for an exclusive interview with the newest Herndon hero, the one representing the class of 1965.

"I would say it was serendipity, in the sense that in the morning I certainly did not anticipate doing anything other than just going over and participating [in the Climb] and being a worker bee,"

Ed explains. "It just so happened that I was in the right place at the right time, and it all happened so quickly. Apparently in three minutes, it was all over."

His most lasting memory?

"I was just happy I was no longer a plebe," he laughs, expressing a very common sentiment.

After graduating from the Academy in 1965, Ed went on to a long and successful career as a naval officer. Any trace of a longing to return to the "farm" was purged from the "boy," and the lure of the sea beckoned in the specific form of the Submarine Service.

"I absolutely loved being at sea. I loved being on board submarines," Ed says fondly of his underwater years. "Serving on board a submarine is like the ultimate boys' night out."

His first assignment was on board the USS *Gurnard*, a nuclear-powered fast-attack submarine with a crew complement of 109, including fourteen officers. While the *Gurnard* and her crew engaged in tricky games of Cold War cat-and-mouse with the Soviet fleet, Ed found numerous opportunities for advancement, and after five years of service he rose through the leadership ranks from a newly assigned officer, fresh from the Academy, to the third officer on board the submarine. "We had very successful special operations. Very successful," Ed says without elaborating, as the exact nature of the *Gurnard*'s mission is not a matter of public record.

Roughly four years into his tour on the *Gurnard*, Ed planned to resign from the Navy, with a view toward attending law school. Those plans were deep-sixed when he found that President Lyndon Johnson had exercised one of the lesser-known presidential powers, unilaterally extending the commissions of most naval officers due to wartime necessities, as not only the Cold War but the Vietnam conflict stretched the capabilities of the entire U.S. military. So when he eventually disembarked from the *Gurnard*, following a brief stop at the wedding altar where he married his sweetheart, Sharon, Ed boarded the ballistic missile submarine

USS *Nathanael Greene* as navigator and operations officer, and another successful tour ensued.

Next, Ed took advantage of a program that allowed qualified officers to take time off from service to further their civilian educations while still remaining "on call" with the Navy. He had applied and been accepted at the business schools of Harvard and the Massachusetts Institute of Technology; but he ended up at Oxford's Christ Church College, where he earned a master's degree in a groundbreaking interdisciplinary program called Philosophy, Politics, and Economics, or "PPE." Ed refers to his time at Oxford as "an intellectual sabbatical," for which he had agreed to extend his service, and he expected the Navy would insist on "payback."

That demand for payback manifested itself promptly, and Ed was assigned to serve as executive officer on board the USS *Tinosa*, a nuclear-powered fast-attack submarine of the *Permit* class, a design that was actually less advanced in its capabilities than the *Sturgeon*-class *Gurnard*. At the time, the *Tinosa* had a spotty record of service and a crew that was riddled with a Murphy's-law book of disciplinary problems, and as her new XO, Ed did his level best to restore the ship and crew to the highest Navy standards.

His efforts were rewarded with an assignment to command the *Benjamin Franklin*–class USS *Kamehameha*, making Ed the first member of his Naval Academy class of 1965 to be tapped as commanding officer on a ballistic missile submarine.

"I was riding high and doing well," says Ed, "and you might say I was well on track in a very rapid successful career in the submarine force." And for a time, he admits that he seriously entertained the notion that he would fulfill the mythical destiny of a Herndon "scaler" and become the first admiral of his class; but flag rank was not in the cards for him.

Over the years, Ed had grown increasingly disillusioned with what he saw as a profound difference between the Navy he expected to serve in after his Academy training and the disturbingly draconian approach to the enforcement of regulations manifest in the newly emerging *nuclear* Navy. In a key turning point, he found himself helpless to prevent the demotion and career trashing of a promising junior officer, for what Ed perceived as nothing more than a benign error of record keeping. Eventually his disillusionment moved him to retire from the Navy in 1985 after a two-decade career.

Twenty years of duty as a submariner seems to have been just the beginning of a fascinating life of service, regularly punctuated by adventure, challenge, and personal achievement. Since his retirement, he has indulged a passion for youth sports coaching that had developed between sea deployments. He also attended night courses at George Mason University and earned a second master's degree, this one in secondary education, a springboard into a new career, teaching science and mathematics to students who doubtlessly peppered him with as many questions about life at sea and on board submarines as about quarks and quadratic equations.

Content in his new career, in the early 1990s he experienced a brush with death in the form of a diagnosis of cardiac sarcoidosis, a heart condition that is rarely less than fatal. Although he had no noticeable symptoms, in 1991 doctors informed him that he was unlikely to survive for more than three years. That projection proved to be extraordinarily accurate, and by 1994 Ed found himself on the short list for his only chance at survival: a heart transplant.

"I was in the hospital for three months, being kept alive, barely," Ed recalls, "and on the ninth of September in '94, I gratefully received the heart of a thirty-three-year-old woman who had been in perfect health but who passed away suddenly from a brain aneurysm. Everything goes great [during the transplant surgery], so thirty-six hours later I'm sitting up in a chair, thinking 'Oh my God, I'm alive.' That lasted for about two hours and then my lungs and kidneys crashed from the cumulative effect of the drug therapy I'd been getting, and I went into a coma that lasted three weeks."

One miraculous recovery and seven months later, Ed was hiking to the bottom of the Grand Canyon with his son and daughter and hiking back to the upper rim the following day, with a new lease on life and profound faith in the capabilities of medical science. Since then, he has been symptom-free, with a healthy vigor he had not felt since the long-ago day that he climbed the chicken wire fence around the Herndon Monument.

Never satisfied to rest on his laurels, and determined to prove himself worthy of a young woman's unfortunate and unanticipated sacrifice, Ed has redoubled his efforts to live a rich and exemplary life. He is the author of several published books, and he travels tirelessly to remote corners of the globe. Nowadays, a typical year in Ed's life includes not only a Grand Canyon hike, but a Tanzanian safari, a Key West fishing expedition, and a casual perusal of Sicily's famed Mount Etna. He faithfully records his observations in regular contributions to a Seattle newspaper and to his blog site *Eyes Right*, where he fearlessly tackles subjects ranging from the controversial to the mundane.

One might quibble with his political viewpoints, or indeed his scathing review of the Tony-winning Broadway show *The Book of Mormon*, but Ed leaves no doubt that his has been a life well-lived.

5

THE LONGEST CLIMB

(or, Four Hours Together by the Bay)

Four years together by the Bay
where Severn joins the tide
Then by Service called away
we're scattered far and wide.
But still when two or three shall meet
and old tales be retold
From low to highest in the Fleet
we'll pledge the Blue and Gold.

—From "Navy Blue and Gold"

STEVE FORTMANN had the best seat in the house.

Or maybe the worst. It depends on your perspective.

Every year, the Herndon Climb draws thousands of spectators. Midshipman families and friends are in town for the various Graduation Week functions, and numerous alumni bring their families on an annual trek to witness the festivities despite the prospect of scarce and exorbitantly priced accommodations and long waits for tables at the finer Annapolis restaurants. Locals, on the other hand, can find no better entertainment bang for their buck, since there is no charge for admission to the event.

Mark Harper, class of 1975, taught engineering and science courses as a civilian professor at the Academy for twenty-five years, on the heels of fourteen years of active duty, and he continues to make his home in the Annapolis area. He has hardly missed a Herndon Climb in the past four decades.

"Over the years, I've had a lot of fun going, and my family comes over to the Yard, and we make a family picnic of it," Mark says, "and we've watched many, many classes do the Herndon Climb."

And every year, after about twenty-three minutes elapsed time for the plebes struggling against the forces of gravity and grease and granite, Mark smiles and celebrates the fact that his class of '75 accomplished the Herndon Climb in one of the fastest, if not *the* fastest times ever recorded.

But when Mark and his family spread a picnic blanket, sipped cool beverages, and munched sandwiches and snacks on that sunny May day in 1995, as the class of '98 approached Herndon, his vantage point was quite different from Steve Fortmann's.

The Harper family picnic was set up on the Yard, maybe fifty yards away from the monument and the mass of plebes attempting to scale it.

But Steve eventually found himself with a unique view of the proceedings. A strategic observation post, something like a crow's-nest on a sailing ship, where he could see the entire Yard, the spectators and participants inclusive.

He was seated, or perched, at the very *top* of the twenty-one-foot granite obelisk.

And he wasn't happy.

By the time Steve finally accomplished the task at hand, removing the Dixie Cup and triumphantly placing the combo cover—his *reason* for seeking that lofty perch at the apex of the monument commemorating the sacrifice of Commander Herndon—more than four hours had passed since the sound of the first cannon at the start of the sweaty ordeal. That year, the annual Harper family picnic went on so long even the ants had their fill and needed a nap.

The class of 1998 holds a different record, one with a dubious distinction: the longest time ever recorded for a Herndon Climb, at four hours and five minutes. A record of futility, perhaps, but the fact that they never gave up and that they ultimately succeeded, asking no quarter or lax interpretation of the ritual's unwritten rules, is a testament to the strength and pride of the entire class, and particularly that of a core group of determined individuals who simply refused to quit.

Examining only the times of the various climbs over the years, one might jump to the conclusion that the '98 plebes were something like the Navy equivalents of Beetle Baileys and Sad Sacks, but the opposite is true. The extraordinary courage of Steve Fortmann and the unflagging support of his classmates exemplify the best of the Academy's ideals and the finest values of the young men and women who made up the Brigade of Midshipmen.

The youngsters of the class of '97 did the plebes no favors that year when they prepared the monument the night before the Climb, slathering on the usual heavy layer of grease; whether they used lard or Crisco vegetable shortening mixed with glycerin no one seems to recall, but the point is irrelevant. What is most relevant is that one or more of the engineering geniuses among the '97 youngsters produced a masterpiece of diabolical creativity.

There are no discrepancies in the inventory records of the Academy Post Office on the day in question at the beginning of Graduation Week. But somehow those geniuses of '97 managed to obtain a supply of nylon-reinforced, heavy-duty strapping tape. You know the kind: not duct tape, which is strong but easily tearable, but the type of packing tape that you find in post offices and shipping/delivery operations. It must be cut with scissors or a cutting tool or with the serrated edge of a handheld or desk-mounted dispenser.

That's the kind of tape they used, to not only secure the Dixie Cup to the top of the monument, but to bury it from view, beneath roll after roll after roll.

Spectators and participants recall that the '98 plebes made a good start on the Climb, exhibiting an uncommon degree of organization and singularity of purpose. They were quite effective in removing much of the grease and, most importantly, getting strong backs against the base of the monument in a tight circle of bodies, linking arms firmly, forming the critical first layer of the human scaffold. In less than thirty minutes, the second and third layers were solidly in place, and long arms were reaching the apex.

And that's when the trouble started.

The plebes who had scaled the pyramid of classmates and were nearest the top quickly realized that the task of removing the Dixie Cup had gone to an extra level of difficulty. Initially, some thought there had been a mistake and the youngsters had failed to place the cap at all. But soon it became apparent what they had really done: the Dixie Cup was there, but it was trapped under an impenetrable thicket of the most durable, toughest-sealing adhesive tape available to the Navy.

Word began to filter down to those in the crush below that this was going to take a while. And if they already felt like mortal versions of Atlas, holding the weight of the Earth on their shoulders and heads, the task had only just begun.

"I was at the bottom," recalls Brian Weaver. "I'm a bigger guy, so I ended up at the bottom. And unfortunately, at the bottom, once you're in there, you're in there. There's not a good way to rotate out."

And although it's true that over the years, many in his position in that ground-level ring, crushed from all sides and ankle-deep in mud, have managed to wriggle their way to freedom and fresh air, Brian stood his ground, testing the limits of his strength and endurance.

"So I stood for the bulk of the four hours, probably three hours and forty-five minutes of that, with people piled on top of me," Brian says, his voice touched with a bit of well-earned pride. "I couldn't see anything of what was going on. I had to get everything from second- and third-hand reports the whole time. Visually? There were a lot of crotches, is what I could see as people climbed past me."

Ryan Dupnick was there too, with a similar view and the same crushing weight on his shoulders.

"I knew my place, for sure," Ryan remembers, with a hearty laugh. "Coming off freshman year, I probably weighed 225 pounds, so I knew. Hey, I'm on the bottom. If I get on anybody's shoulders . . . no. No, I'm gonna get on the bottom and have people stand on me. I'm not going to try and scamper to the top or do this or that other thing. I'm going to put myself where I'm the most useful. And link arms, just as strong and tightly as I can."

The hours passed slowly. By all reports it was a hot, unusually humid day on the Yard; endurable enough for the spectators, but something akin to the Black Hole of Calcutta for the plebes of the class of 1998 in the pile of straining, exhausted bodies.

"After two hours, I had scaled the human pyramid at least three or four times that I can remember, and I had fallen back down," says Rex Trudell of his experience with the ordeal. "I remember a couple of times I just had to walk away. I was just thinking, 'this is ridiculous.' It was still almost another two hours before he [Fortmann] finally got up there. He had somehow pulled his way to the top, and he was literally sitting on top of the monument."

It wasn't his first time near the top; just like Trudell, Fortmann was one of the slippery bodies that tumbled down, in one after another partial collapse of the rickety human scaffold. He was not destined to reach his exclusive perch until his second ascent.

"After I don't know how many hours, I was able to get up there the first time, and I fell hard," Fortmann recalls. "It was a pretty, pretty long fall from the top. I just recently looked at a video of that. I just think 'Wow! That's a bad fall!' I remember they made me go over and get checked out."

After a cursory evaluation by an overworked medic, he was deemed fit for duty, and soon he was back in the Climb.

One thing Navy SEAL Steve Fortmann does not have is a fear of heights or a fear of much at all; that is part of a frogman's job description.

Something he *does* have is a set of strong, gleaming-white choppers. And he put them to good use against a tough, determined foe: that impregnable matrix of packing tape between himself and the elusive Dixie Cup.

It doesn't seem fair. The plebes are not allowed to use tools, ropes, cargo nets, or anything but their bodies and their shirts and shorts; for women, the dress code accommodates bathing suits. In earlier years, shoes could be worn but nowadays those are contraband, too. But the youngsters of '97, in using that huge quantity of tape—something that might be deemed to be of questionable permissibility—may have justified another exception to the rules.

Various options were pondered and discussed on the reviewing stand and among the spectators as Steve, with no other tools handy, attacked the layers of packing tape with his teeth. He would have taken any number of demerits in exchange for a pocket knife, a razor blade, or even a pair of needle nose pliers. He slipped a bit from his perch, once or twice, holding on by just his bicuspids and molars, using the tape itself, as a clever warrior might use an enemy's own weapon against him.

"I'm not a rock climber, but at that point, I was good at doing pull-ups," he points out unremarkably. "I was able to basically do a one-arm pull-up, with my fingers inside the tape. I think if the tape wasn't there, and so anchored to the top, it wouldn't have been possible to pull up on top like that."

As Steve explains it, multiple strips of tape ran up the sides and over the top of the monument, strapping the hat to the apex. Then, several horizontal layers were wrapped all the way around, securing the entire construct of vertical bands and giving him a satisfactorily secure handhold.

"At some point, we heard that the superintendent [Adm. Charles Larson] had sent word, to just put the cap up there, and be done with it," Brian Weaver recalls.

But as any parent knows, sometimes your kids are struggling with some or other task and you want to help them just a little bit, and they say, "No! No! I wanna do it myself!"

The class of 1998 was as stubborn as any child.

"The Naval Academy itself is about challenges," says Brian. "Overcoming challenges and persevering through tough times and preparing you for life. And Herndon was all of that in a very . . . " he paused to chuckle at the irony, "well, in a *relatively* short time."

Meanwhile, Rex Trudell had discovered something. And perhaps it was divine intuition or the mother of invention that showed him the value of the thing he discovered.

Others may have had the same idea at around the same time. But Rex was in a unique position, just below the dangling feet of Steve Fortmann, one hand against the monument and the other hand holding a midshipman's combo cover. That cover and others like it that were available to top the monument were the only other tools they had.

More to the point, and most relevant to the task at hand, was the insignia attached to that cover.

A nice little hunk of brass-plated aluminum, in the shape and image of a "fouled anchor."

Rex noticed that the flukes of the anchor insignia were relatively sharp, and at each tip the bill came to a fine point. And, he reasoned, if the flukes of an admiralty-type anchor are designed to dig into the sea bed and hold a seagoing vessel in position, might these much smaller flukes be used to tear through a few layers of nylon-reinforced packing tape? If they weren't quite as sharp as Rex would have liked, certainly they were sharper, and considerably harder, than his classmate's tooth enamel or greasy fingernails.

So, balancing himself precariously on the hard but slippery shoulders of a classmate, Rex tore the insignia from the cover, and somehow managed to toss it to Steve.

The annals of the sea are replete with tales of how a strong anchor and a proper anchorage saved many a ship from foundering during deadly storms and other perils. And it has been suggested that this one small anchor-shaped insignia, designed for decorative rather than practical purposes, helped save the class of '98.

With the makeshift knife in hand, Fortmann greatly expedited his progress. And soon, the longest Herndon Climb in recorded history came to a spectacular end. Cheers echoed across the yard.

The triumphant cry finally went up:

"Plebes . . .

" . . . No . . .

" . . . More!"

And the celebration began, belatedly, but with an extra measure of satisfaction for the class of '98, as they realized that they had overcome one of the most difficult challenges they would face in the course of their lives.

At some point during the epic four-hour ordeal, nervous and knowing glances began circulating among the class of '97 youngsters who had prepared the monument. Those looks spoke articulately, and it was impossible to misinterpret what they meant: "Maybe we went a *little* too far . . ."

And as the afternoon wore on, threatening to turn into evening, family members and Ring Dance dates began to pile up, in a manner much different from the pyramid of plebes: those unfortunates were stranded at Baltimore/Washington International and Dulles and Ronald Reagan airports, while their chauffeurs, the upperclassmen among the Brigade of Midshipmen, eagerly and impatiently awaited the moment when they would finally be given liberty. But as a matter of military custom and tradition, liberty could not and would not be granted until the mission was accomplished, until the monument was officially scaled and topped and all the ceremonial whoop-de-do was good and done.

All they could do was wait. While friends and family members endured mild privations and emptied vending machines at BWI and the other local airports, the midshipmen were obligated to remain at their duty stations as the Climb wore on.

Gradually, the seniors and juniors began to add their own, silent vocabulary to the growing panoply of sideways glances and subtle eyerolls and head shakes that spoke directly to the youngsters. "Yes. You went too far," was the obvious consensus of opinion among the upperclassmen.

Eventually the monument was topped and liberty was granted, to everyone's great relief. And, as one might expect, a certain measure of revenge was ultimately extracted.

At least one of the suspected ringleaders among the '97 youngsters was targeted. Perhaps she was a scapegoat, an innocent victim of the payback; but regardless of her role in the packing-tape scheme, "Sally Smart" paid a certain price for the dirty pool played by her classmates. One or more unidentified individuals broke into her dormitory room, piled most of her possessions on her mattress, rolled up that mattress into a secure cocoon, and wrapped in multiple layers of the same notoriously strong, reinforced tape.

"Maybe she was late to noon formation, but I doubt it took her longer than four hours and five minutes to get her stuff," says one of the many suspected culprits.

If Steve Fortmann knows who did it, he isn't saying. And he bears no trace of animosity toward Sally Smart.

"She executed her job with full diligence, as far as I'm concerned," he says with a laugh.

Flash forward: Graduation Day 1998. Three years have passed, and not one of the seniors has forgotten about the day, and the moment, when they ceased to be plebes.

The featured speaker at the commencement ceremony that day was President Bill Clinton. He began with a few jokes, about how the apple dumplings in King Hall were called "cannonballs" because of their size and heft, and that the popular film of the day, *Titanic*, was about a ship that sank. Those jokes mostly fell flat, drawing scattered laughs, but then Clinton, or his speechwriters, couldn't resist the temptation to poke a little fun at the graduates about their, shall we say, *noteworthy* performance in the Herndon Climb three years earlier.

"I will try to be relatively brief today," said the commander-in-chief. "I was given only one instruction: I should not take as long as your class took to scale Herndon Monument. At four hours and five minutes, the slowest time in recorded history, I have a lot of leeway." This remark was greeted by gales of laughter and applause by the graduates, who clearly had come around to embrace their final adventure as plebes so long ago.

"But you have more than made up for it," Clinton continued, appropriately changing course. "You have done great things. Succeeding in a rigorous academic environment, trained to be superb officers. In basketball, you made it to the NCAA championship tournament for the second year in a row."

Then he came to the pièce de résistance: "In football, you defeated Army last year!"

Those with a semblance of familiarity with Naval Academy culture and tradition will not be surprised to learn that this was the biggest applause line of the president's entire address.

The class of 1998 not only embraced their distinction as the class with the longest recorded Herndon time, they made it their motto.

The slogan "Four Hours by the Bay" was immortalized and emblazoned on caps and T-shirts and beer tankards and other commemorative objects.

And a new verse was composed and popularized for the classic "Navy Blue and Gold" fight song.

But few can recall the complete set of lyrics for that verse, and those lyrics that are best remembered include language that is salty as the mighty Atlantic.

6

THE HEIGHT OF GLORY, PART 2

(or, Slip-Sliding Away)

> Never tell people how to do things. Tell them what to do, and they will surprise you with their ingenuity.
>
> —George S. Patton

For the first time since you set foot on the grounds of the Academy last summer, you see the eastern face of the monument. There, about a third of the way up the side, you see the name of Herndon standing out in stark relief on the dark, hard granite.

It is hardly in a clean, pristine state at the moment. Owing to the diligent early-morning efforts of the second-year midshipmen of 1st Company, the monument now sports a heavy coating of grease. Soon you will learn just how much this increases the difficulty of the task, and you knew to expect it. A few days ago, the talk among you and your roommates turned to the Climb, the little bit you each knew about it, and what kind of "grease" you would encounter. "Probably lard, or Crisco, maybe butter," is what you've heard. Lard or Crisco? What's the difference, anyway?

One of your mates, raised around a family baking business in Skokie, Illinois, has a ready answer. "Lard is rendered pig fat. Crisco is not." Sounds pretty disgusting, but you learn that many professional cooks and bakers prefer lard to Crisco or other commercially available vegetable-based shortenings because it results in a pie crust or pastry that is lighter, flakier, and, believe it or not, more delectable than a similar product made with vegetarian-friendly ingredients. You added that bit of gastronomic intelligence to your ever-growing store of knowledge, and eventually you realize that the distinction may be important to a pastry chef, but to a plebe in the throes of a Herndon Climb, there is no difference. Lard or Crisco, it all makes a slippery mess of things, and it will infinitely complicate your efforts.

You have finally reached the end of the brick pathway and now you are standing on the grass that your plebe feet have never been permitted to touch. For just a moment, you pause to contemplate the sight before you and your place in this huge, chaotic, extraordinarily important event, so rich in history and tradition stretching back to before the 1940s, with an ever-evolving set of rules. You understand the goal, which has never really changed: get one person up to the top of the monument, remove the Dixie Cup if it hasn't already been knocked from its perch, and put an officer's cover in its place.

But as the wicked witch observed in *The Wizard of Oz*, the question is not what you will do, but *how* to do it, and there is only one way. Unless one of your classmates has the ability to fly without an enchanted broomstick, you have no option but to use the combined strength of your own bodies. You must form a human pyramid, a scaffolding to surround and surmount the unforgiving granite, a ladder of muscle and bone and blood and sweat and perseverance, for one of your classmates to climb and to reach the perilous height that seems so far aloft.

A couple of days ago you received an e-mail from one of your classmates you have never met since he is part of a different company than yours. It said, basically, that a group of class representatives has held a surreptitious meeting and developed a plan to conquer Herndon in record time.

According to this plan, you will disregard your company affiliation and organize yourselves in a rough order based on your size and strength. The largest and strongest among you, the football linemen and shot-putters and their equivalents, will report for bottom-level duty and surround the base of the monument. Those carrier-class bodies will link arms, tightly and closely, and encircle the monument. In turn, they will boost other, slightly smaller classmates on their shoulders, to form a second layer of support. And so forth until the "monkeys," the best and most agile climbers—who have demonstrated their ability to scamper up cargo nets and easily surmount the walls of the numerous, famously challenging Naval Academy obstacle courses that have been part of your training regimen since day one—will attempt to reach the granite apex. And you must make sure that when they get there, they are not empty-handed or bare-headed, as they must carry the all-important combo cover that will cap the monument; and then, it's all over but the cheers and chants of "Plebes no more!"

But a few hours later you received another e-mail from another classmate in another company. He describes a similarly surreptitious meeting where it was decided that the best strategy was to tie together a dozen of those discarded, sweat-stained, blue-rim T-shirts into a line that could be used in the manner of a climbing harness. This harness, and others like it, would help boost the climbers and reduce the pressing weight burdening the lower layers; and with such an improvised tool, properly constructed and deployed, one person should be able to quickly reach the top of the monument. Maybe you thought this plan was dubious, because it was tried just a couple of years ago and the result was one of the slowest times in recent history.

No matter, because a few hours later came yet another e-mail describing a completely different but equally secretive meeting at which a completely different "plan" was developed and agreed upon. Then came another. And another. And then you stopped reading e-mails.

Each of these plans, when confronted by the reality of the Climb, would ultimately bear as much relevance to the task as the best-laid plans of the generations of house mice that populate the ancient, towering rafters of Bancroft Hall.

You take a moment to contemplate your own role in the unfolding, chaotic yet purposeful drama. Any shred of organization has been discarded, so apparently you are a "free agent": How exactly you participate in this group effort is entirely up to you. You notice that some of your larger classmates are already carrying smaller bodies piggyback-style, attempting to boost them up the side of the monument. You aren't particularly big and strong nor small and wiry; maybe you are slightly above the average size among Navy personnel. Some of your friends and family back home, when you told them that you were headed for the Academy, wondered, "How is someone your size going to maneuver through doorways, or hatchways or whatever they're called, on a submarine or one of those small ships? What are they, destroyers?" Most civilians, you realize, are familiar with the small size of a destroyer, relative to other Navy vessels, from the famous peg-board game of "Battleship."

So, you reason that the best place for you is not at the bottom, where bulging muscles and steel-strong spinal columnae are de rigeur. Nor should you attempt to reach the top; that assignment is best taken by the long-armed beanpoles and the smallish, compact powerpacks, the volleyball players and the gymnasts respectively. So as your classmates swarm the monument and begin to pile on top of each other, like a mass of singular-minded zombies in a scene out of *World War Z*, you decide that your best place is somewhere in the middle.

The air is filled with projectiles. Not bullets or shells but balled-up socks and T-shirts. You feel inspired; if the monument was a United States flag and there was a red glare of rockets hanging about, such a sight might have been what inspired Francis Scott Key to write his fairly well-known poem a mere thirty miles to the north. The T-shirts are being put to good use as washrags to combat the half-inch thickness of grease coating the monument; helpful, but an imperfect solution, as a slick surface is left behind and only a high-pressure washing will ultimately restore Herndon to its clean, pristine state. Socks are being hurled in the general direction of the Dixie Cup, with the intention of knocking it from its perch, and mostly with the futility of players at a rigged carnival midway game.

You notice that the "youngsters," or sophomores, who prepared the monument have left a variety of messages for you, written by their own fingers, in the slathered-on layer of grease. These messages range from the mocking to the sarcastic: their own class year, not yours; a taunting "Brace up, Plebes!" and a creatively inspired "Halfway to the Top!" which is inscribed not halfway, but barely a third of the distance to the apex of the monument.

A T-shirt strikes the back of your head as you see an opening and climb into it. You haul yourself up with an assist from a classmate who presses upward on your tailbone, and presently you are standing on the broad shoulders of one of the strongmen who have bravely taken bottom-level duty. It's a slippery footing at best; perhaps the grease is being removed to some extent but there are traces of it everywhere, and it spreads to every surface of bare skin it touches, like a viral contaminant. But now you are dedicated to the task, and you link arms as tightly as possible with the mates on your right and left. You press your back against the monument and gaze skyward. The taunting Dixie Cup seems as far away as the stars and you wonder, briefly, what you have gotten yourself into. You may be stuck here for an hour, or two, or more. There is precious little opportunity to rotate out, and although your classmates are willing and able to relieve you of duty, there is no practical way to do it. Just like on Induction Day, when you raised your right hand and swore the oath to support and defend the Constitution of the United States against all enemies, foreign and domestic, you are committed.

You quickly realize that tightness and closeness are the keys to establishing a strong base at each layer in the human scaffold of plebes. The slightest amount of play between adjacent shoulders will inevitably introduce a wobble, no matter how much pride you and your classmates take in the strength of the muscles of your arms, and it will be amplified and multiplied and distributed to the next level above, vastly increasing the danger of collapse of the entire carefully constructed structure. The beleaguered fellows below you are packed much more tightly than canned sardines, their backs crushed against the unyielding granite, the collected force of dozens of bodies pressing against them in the surrounding crowd.

Now, your fortitude is about to be tested as a third level is beginning to take shape. It's time for you to be crushed from all sides and from above as classmates are pushed up, often slipping and falling, and eventually you find one foot on your right shoulder and a different persons on your left. And you reach just a little bit deeper into your reserves of energy and endurance.

The activity around you is frenzied, and the stability of the scaffolding you have constructed from your own bodies has become suspect; you are like a house of cards in a gentle breeze that is steadily intensifying. You have managed to adjust your position, ever so slightly, so that the load supported by the muscles of your back and legs is evenly and properly distributed. Your feet are well-planted, rolled back to take the weight on your heels rather than the arches, shifting it off the slender muscles of the front legs to the stronger load-bearing hamstrings. A year ago, you couldn't have done what you are doing now; endless hours spent in the gym and on the exercise field throughout plebe year are now paying off. Not only has your strength approximately doubled, but you are also benefiting from an increased understanding of kinesiology and leverage and physics.

Around you, everyone is shouting or grunting or groaning. Above this racket you hear the cheers of the spectators, not in a steady roar but an alternating sequence of upswelling, rising-pitch excitement, followed by agonizing decline, the familiar group call of a sports stadium full of fans whose team is coming tantalizingly close to victory, then falling just short.

You have no way to know what is happening on the other three of the four sides of the monument, and your view of the action on your own side, the one facing Zimmerman Bandstand, is also greatly obscured. But you infer from the temporary stability of the ring of classmates above you that climbers must be getting close to the top. Assuming three stacks of people, averaging a little less than five feet per stack measured from foot to shoulder, means about fourteen feet, just seven feet from the apex, a seemingly reachable distance. But anyone attempting to stand upon that third level will find dangerously unsteady footing. Already you have seen through the daylight spaces that open and close around you that many people are climbing and falling, and they seem to be increasing in number.

Pressure waves undulate among your closely packed classmates, now so tightly intermingled that the strength of the encircling ring of bodies sometimes lifts you slightly upward, not enough to raise you completely off the shoulders below, but noticeably. This must mean an imbalance, or weakness, somewhere in the structure. You have a vague premonition of danger.

And then, disaster strikes . . .

(To be continued . . .)

7

THE PRIDE OF THE BIANCHIS

(or, A Tragedy, a Tattoo, and a Triumph)

> Greater love hath no man than this, that
> a man lay down his life for his friends.
>
> —John 15:13, KJV

CHRIS BIANCHI was just six years old when his father died.

By any standard, that's much too young for a boy to lose his father. But it's also old enough for that boy to retain vivid memories of the man and to recall the terrible day he heard the tragic news, and to keep firmly in mind for the rest of his own natural life the profound example of honor and courage, the legacy of service and sacrifice that his dad bestowed on him.

Chris, a member of the United States Naval Academy's class of 2019, is determined to honor that legacy.

On the sixteenth of July 2003, Cdr. Kevin Bianchi was the senior officer of a four-man crew, copiloting on a routine training mission near Naval Air Station Sigonella, when his MH-53E Sea Dragon crashed after a fire broke out in one of the heavy-lift cargo helicopter's twin engines.

There were no survivors.

Lt. Peter Ober, PO1 Brian Gibson, and PO3 Samuel Cox were also killed, along with Commander Bianchi, a 1985 graduate of the Naval Academy.

Among Bianchi's personal effects recovered at the crash site, on the embankment of a pond just outside the town of Palagonia, Sicily, was his wedding band, which he always kept in his back pocket on the occasions that it wasn't on his finger.

Today, his son Chris wears his late father's ring on a chain around his neck, along with his mother's similarly unadorned golden band, which is a bit smaller and fits just inside his father's. On the rare occasions that he removes that precious chain, he keeps it in his desk drawer, very close at hand.

Under some circumstances, anything worn around the neck, from chains and necklaces to silk ties and dogtags, can present a strangulation hazard, so on the day in 2016 when Chris and his plebe classmates eagerly commenced the Herndon Climb, that chain was safely tucked away in his room in Bancroft Hall.

After one hour and twelve minutes had elapsed, the Climb reached its conclusion, and the human scaffold of plebes tumbled into chaotic joyful collapse immediately after one climber made his way to the top of the slippery pile and triumphantly placed an officer's combo cover at the apex of the monument.

That plebe who achieved that distinction representing his classmates of '19 was Chris Bianchi.

For Chris, the exuberant feeling of reaching the official end of plebe year and playing such a key role in his class' success at Herndon were the diametric opposite of the emotions he felt on that other day when he was just six years old, living with his family at Sigonella, and heard the terrible news that his father was gone.

"I remember a lot about that day, which is interesting because most of my memories of my father come from stories that all of his friends that are still in our lives talk about, which is a lot of people," Chris explained during a short break in his extraordinarily busy daily schedule, just a few months from his highly anticipated Academy graduation date. "And then all the videos and pictures and especially, of course, from my mom and my older brother. But that day was vivid. I have certain memories that will never go away."

The Bianchis lived a typical military family's itinerant lifestyle, following Kevin's career as a Navy helicopter pilot and mission commander, from Virginia Beach to San Diego, to Virginia Beach again, and finally to Sicily, from where Chris retains his earliest memories.

He recalls, without relying on photographic evidence, that his family could see Mount Etna from the driveway of their home at the airbase, and that they often found a soft layer of ash underfoot, to accompany the view, whenever the famous and still-active volcano had been in a cranky mood and the prevailing winds were blowing in their direction.

That driveway was also the site of rope-jumping contests and games of wiffle ball, figuring prominently in some of the happiest memories Chris has of his father.

He also vividly recalls his father's fortieth birthday celebration, when he and his brother argued over which one of them would have the honor of lighting the birthday candles, "or, some other stupid thing, like brothers fight about," laughs Chris.

Later, he recalls that he and his brother, Kevin Jr., age nine, and younger sister, Julia, age three, stayed at the nearby home of a babysitter, while Mom and Dad went out on the town, continuing the birthday festivities.

The following day started out normally enough. Commander Bianchi left the house early to commence his regular duty shift with Heavy Helicopter Combat Support Squadron 4. He wasn't scheduled to fly that day, but according to Chris, he loved flying so much that he asserted his executive officer privilege and put himself on the flight schedule.

He never returned home.

"We were doing some kind of family activity," Chris recalls. "I want to say we were at the pool, or at the movies. Something normal like that. And my mom got the call to come over to the base." So far, it was nothing out of the ordinary. Chris and his siblings had no reason to suspect that anything odd was going on when they arrived at the squadron, with no further news or reason as to why the family had been summoned. But the three Bianchi children were promptly taken to the game room, to be served apple juice and play foosball with flight-suited personnel—his father's squadron-mates and friends—while Mom was escorted to the church.

Chris' memory extends to more than foosball and boxed apple juice. "Looking back for me, being in the military now, I think about how hard it must have been for those guys to hold it together," says Chris, "because my dad was one of their really good friends. They were just keeping us busy and entertained while [the base chaplain] broke the news to my mom. And they're looking at me, my brother, and my sister knowing that we had just lost our dad. Kudos to them, because that's tough, to have to hold it together for the kids and play with us like nothing was wrong, but knowing this pretty deep tragedy."

After some time, the children were finally taken to join their mother.

"We walked into the church and she was on her knees, hysterical, very upset," Chris remembers. "So we ran up to her, and at that point [the chaplain] gave us some space. And we just asked her, 'what's wrong, Mommy? Why are you crying?'"

Still brimming with the typical innocence of children, they couldn't imagine the worst. "Is it Daddy or something?" one of them finally asked their grieving mother. Chris doesn't even recall which of them it was, he just knows one of the children blurted out that prescient question.

His brother was older, perhaps less naïve, and his younger sister even more innocent than himself, but Chris remembers thinking that it might have been something as terrible as a broken leg.

But finally his mother managed to choke through the tears. "Daddy's gone."

For Chris, the rest of the day is largely a blur.

But the tragedy could not erase the childlike innocence. That night, as bedtime loomed, young Chris tugged on his mother's dress and asked hopefully, "Mommy, since Daddy died, can I stay up late tonight?"

The widowed Rita Barrie Bianchi—who today describes her mostly successful efforts at always being "a very calm Navy wife" as a coping mechanism to deal with the ever-present knowledge of the hazards her husband faced on a daily basis—carried on despite her loss, raising her three children on her own. She goes by Barrie, and she credits her extended family, and Kevin's, as well as the entire support structure of Navy friends and acquaintances with helping her through the toughest times.

"I was born in a wonderful family, and then I married into a wonderful family," she says, "so I got double-lucky. Double blessed."

One day, Barrie was pondering the impracticality and futility of attempting to straighten the bedroom of her sixteen-year-old son Chris, when she found something that surprised her. It was a large, folded sheet of drawing paper on which the boy had painstakingly drawn an image, the likes of which she had never expected to come from the hand of her own son. The paper wasn't particularly well-hidden, not safe from a thorough maternal shakedown, but not at all in plain sight. She was certain that Chris hadn't meant for her to find it.

Barrie sat on the edge of the bed and looked at it. It was nothing unusual; certainly, lots of boys draw such things. It was nothing to be ashamed about at all. She did not think, "I'll have a talk with him, and nip this thing in the bud!" Nor did she put it back where she found it and pretend she had never seen it.

Instead, she examined the drawing, very closely. And she thought, "Wow. This is pretty special."

Chris had carefully designed and meticulously crafted a sketch, which Barrie guessed was intended to be a tattoo, although she had no idea if it was purely a fantasy or if Chris entertained notions of turning it into a reality. And it wasn't an image of a flaming skull or the logo of a heavy-metal band or any such triviality.

It was a tribute to his father.

Among a fairly cluttered composition of distinct images, typical of an ambitious tattoo design, Barrie immediately recognized the Bible verse, written in a flowing, florid script, as John 15:13—"Greater Love Has No One than This, to Lay Down One's Life for One's Friends"—although Chris had deleted the possessive apostrophes in his own artistic rendition of the quote. She recognized the slogan "God Family Country" and a carefully drawn "fouled anchor," the iconic symbol of the United States Naval Academy. She wasn't surprised to see the name Kevin Bianchi and the dates 7-15-63 and 7-16-03, the dates of his birth and death, dates Barrie knew as well as she knew her own birthday. Finally, she noticed one small and easily overlooked detail at the lower left of the drawing, and it touched her deeply and personally.

Attached to the free end of the anchor chain were two rings, plain and unadorned; and one was small enough to fit just inside the other.

One short conversation later, and Barrie understood that Chris was quite serious about the tattoo, and since he was more than a year shy of his eighteenth birthday, the age of legal consent, he needed her written permission for the bodywork to properly begin. She promptly gave her parental consent, and a tattoo artist who was a former SEAL and a family friend stepped forward to do the work gratis.

She was determined to take a positive view of the process and the ensuing result that would be indelibly etched upon the skin of her second-born son, although she knew of no members of either one of her families who had a tattoo, including her late husband. Nevertheless, Barrie was surprised when she saw the finished product.

"I had no idea how big it was really going to be," she laughs, "because he's not a big guy. He's short like his dad was."

True, at five foot three Chris presents a smallish canvas for a tattoo artist, but the finished piece covers almost every inch of available real estate across his back, stretching from just above the trim waistline to the rolling expanse of rock-hard trapezius muscles between the shoulder blades.

"It's all symbolic of what Chris felt from his heart, and then, obviously, his dad's priorities," Barrie observes.

When Chris expressed a firm desire to apply to the Naval Academy, the fresh tattoo had healed over, the skin was no longer raw and painfully sensitive to the touch, and Barrie was supportive, and unsurprised, by her son's decision.

"I felt nothing but excitement. I was so happy that he had the focus and direction to follow that path," Barrie says. But still, on separate occasions both she and his uncle Jimmy, Kevin's eldest brother and USNA class of '81, engaged Chris in heart-to-heart talks to confront any lingering doubts that he was following his own dream and *his* path, not trying to follow his father.

All those doubts were erased as Chris unambiguously stated his passionate commitment to a Navy career, to begin, hopefully, at the Academy. His father's influence was certainly a factor in his decision, but the choice was his own.

"By junior year of high school, after a lifetime of meeting people that knew my dad and the impact he had on them, I decided that if he went to the Academy and his life was awesome while it lasted, that it could be a good choice for me too," he explains.

Barrie supported her son's decision, up to a certain point.

"I told her, this [the Academy] is where I want to go, and I want to fly," says Chris. "She didn't like that. She was okay with me wanting to come here, as long as I didn't fly helicopters. And I can't blame her."

Neither Barrie nor her three children ever met Kevin's older brother Robert, a 1983 graduate of the Naval Academy. Like his brother James before him, Robert was a highly recruited lacrosse player coming out of Columbia High School in Maplewood, New Jersey. Three times he was named all-state and twice all-American at the high school level. In the small but rabidly enthusiastic local lacrosse community in New Jersey and environs, he is a legend: a 1997 inductee of the New Jersey Lacrosse Hall of Fame and named "Lacrosse Player of the Century for New Jersey" by no less an authority than the *Newark Star-Ledger* newspaper. Naturally, during his Naval Academy years, lacrosse was his sport of choice, and at that sport he excelled. After commissioning, "Bob" Bianchi, whom the kids refer to as "Uncle Bobby," completed flight school and, just as his younger brother Kevin would later do, became a helicopter pilot.

On the 23rd of March 1987, Lt. Robert "Bob" Bianchi and three crewmates were killed when their HH-46 Sea Knight transport helicopter, on an assignment to serve the USS *Mercy* hospital ship that had crashed near Cubi Point Naval Air Station in the Philippines.

Bob had been scheduled to serve as a groomsman in the wedding ceremony of a longtime friend; after he died, the couple asked his brother Kevin to stand in for him. At a prewedding soiree at McGarvey's, a classic Annapolis restaurant and watering hole, Kevin first met Barrie, a cousin of the bride. They dated for a while, and one night they found themselves back at McGarvey's, where Kevin asked for her hand; on January 26, 1991, they were married.

The Bianchis welcomed two boys, Kevin Jr. and Chris, who arrived in succession, to be joined by a precious little sister, Julia, or "Jules," whom the family adopted from her native China. As Chris was beginning his career as a commissioned officer, Jules entered the Academy herself in 2019, proudly joining the class of 2023. Kevin Jr. has also embarked on a career path that is anything but humdrum: he's a mixed-martial-arts fighter.

Barrie was well aware of Bob Bianchi's fate when she married Kevin, but she kept the fact mostly out of mind as they raised the children and moved halfway around the world and as he spent a fair number of his waking hours flying and working on board the same big MH and HH cargo helicopters as his older brother had. In a certain respect, she drew a peculiar comfort from the knowledge, thinking that the odds were against the same tragedy striking a family twice.

"I didn't live with fear or anxiety, ever," Barrie says. "I thought that fate was on my side, sort of. I didn't think statistically. I thought I was covered."

Her son respected Barrie's single, insistent stance that he pursue any specialty in the Navy except helicopter piloting. So instead of flying planes or helicopters or anything at all, Chris elected to specialize in explosive ordnance disposal, a career that is hardly conducive to restful sleep for a Navy mom.

"I think being a [Navy] mother is going to be a whole different ball game. It's going to push me a bit further than it did being a Navy wife," Barrie says. "It's going to take much more emotional endurance and strength. It's going to be a challenge, for sure. I joked to Chris, 'why don't you just . . . couldn't you just fly a drone? Like, in a room?' I turn to humor sometimes. But I'm not ashamed to say it really scares me. I'll support him a hundred percent, and I'll just have to keep my prayers up. It's going to push me to really dig deep in my faith."

The fear and anxiety that she kept at arm's length while Kevin was alive are now more frequent, if unwelcome, emotions.

"And then I tell myself, when I get anxious, look at the world. Sadly, people go to school, they go to the mall, they go to church, and terrible things can happen," Barrie explains. "These are among the mom talks I have with myself when I get anxious about the path Chris is choosing."

The term "sea trials" in the Navy refers to a newly constructed and newly christened vessel's first cruise, including an intensive shakedown and testing of all pertinent systems to ensure that the ship is not only seaworthy but ready for a battle situation, the ultimate test of a warship and its crew.

In 1998, following Superintendent Larson's orders to find ways to intensify the rigors of plebe year, the Naval Academy adopted the name Sea Trials for a new program that would become an important tradition. The newly established Sea Trials were inspired by the Marine Corps' Crucible, where recruits nearing the completion of boot camp perform a grueling, forty-mile march over a period of fifty-four hours, through rugged wooded country with no breaks for birdwatching or campfire songs or s'mores toasting.

The Sea Trials bear a certain equivalence to the Navy's own Battle Stations, a twelve-hour series of exercises that complete the training for enlisted personnel at recruit training, or boot camp as it is more commonly called. In Battle Stations, the recruits are subjected to a variety of physical and mental tests and are placed in team-building situations on board a mockup of a Navy vessel to evaluate the effectiveness of their training in basic shipboard functions, triage, and emergency response and communication, and to evaluate their ability to perform in high-pressure situations. Needless to say, the twelve-hour Battle Stations exercise demands the highest level of endurance and concentration. At the conclusion of Battle Stations, the recruits are officially "accepted" into the Navy, represented by the presentation of a cap inscribed with the word "Navy," replacing the daily cap worn during boot camp, which simply says "Recruit." Unlike Academy plebes on Induction Day, who have already achieved some record of academic success at the high school level and who have typically participated at a high level in sports or other significant extracurricular activities, many young men and women who enlist in the Navy come from difficult circumstances, broken homes, and economic disadvantage. So when they don their symbolic Navy caps, for some of them it is the single proudest moment of their young lives, an occasion marked by overwhelming emotion, comparable to a plebe class' final conquest of the Herndon Climb.

The Academy's version of Sea Trials takes place shortly before Commissioning Week. The long day begins at approximately 0300 hours, when upperclassmen bust open the doors of the plebes' rooms to roust them noisily from their bunks. The freshmen are organized by companies, battalions, and squads, and for the next fourteen hours they are required to perform a variety of taxing physical challenges, designed to prove to themselves and the rest of the Brigade of Midshipmen just how far they have progressed with their training since Induction Day, which will now seem like ancient history.

At thirty-six separate stations, each supervised by upperclassmen, the plebes perform endless push-ups and pull-ups; they run through challenging obstacle courses and back and forth across the Naval Academy Bridge for a distance of three miles; and they pummel each other in one-on-one battles with padded pugil sticks. Conventional sit-ups are no challenge for their newly toned bodies, so the Sea Trials require squads to lie in the chilly waters along the bank of the Severn River, lifting a telephone pole with each newly agonizing flex of the abdominal muscles. Fireman's-carrying and four-legged races, staple activities of family summer picnics, are no picnic in the Sea Trials version, coming on the heels of hours of demanding and exhausting physical punishment, where no quarter is expected or given.

Many of the events, such as the foot races and pugil-stick contests, are timed or scored, creating a competition among the companies. At the end of the trials, the results are tallied and the company with the best overall score is announced as the winner and is named the "Iron Company" for that year. It's a title that carries more than just bragging rights: the Iron Company is bestowed with the singular honor of leading the charge on the Herndon Monument.

Chris Bianchi points with pride to his 6th Company's strong performance in the 2016 Sea Trials, placing fourth out of thirty companies, as one of the fondest memories of his years at the Academy.

And careful observers of the Herndon Climb have noticed that the Iron Company plebes who take their place at the head of the line might find the honor to be a dubious one, like being part of the initial assault wave in the D-day invasion of occupied France. Just as casualties were highest among those who were first ashore at Normandy, taking the brunt of the arrayed German defenses, suffering huge losses but smoothing the way for the ultimate success of later waves of attackers, the first plebes to approach the Herndon Monument face a towering and unconquerable smooth-sided obelisk, and if they wish to participate in the Climb from the first cannon shot until the moment of final triumph, their only choice is to throw their strong backs against the greasy granite and begin to form that all-important first layer of the human scaffold.

So Chris and his classmates in 6th Company were in perhaps a better position than the members of the Iron Company, if any of them entertained any notion of being the first to reach Herndon's apex. But from his own point of view, Chris largely dismisses the notion that his company's place in line gave them a leg up, so to speak, on other would-be "scalers."

"I don't think that has any matter to it," Chris says of his potential strategic advantage on the day of the class of 2019's Climb.

Electronic communications technology, which was obviously not available to earlier generations of climbers, has been used in recent decades as plebes examine the history of recorded times and attempt to improve upon them, with decidedly mixed results.

"There are always people that e-mail, on the day of Herndon, their ideas about strategies," Chris observes. "I'm sure that if people actually came together, there's a way to do [the Climb] super fast, and it can be done in like three minutes. But it's just madness when you get out there."

"Everybody, even this year, they always talked about how they—" here Chris lowers his voice to a *Mission: Impossible*–style secretive, conspiratorial whisper, "—they say, 'hey, we met. And we've got a plan.' You see these plans in effect and they almost pan out. Sometimes I think eventually there's going to be a plan that works, and it's going to happen fast."

If the plebes of the class of 2019 had agreed on some sort of well-organized plan, Chris didn't hear about it.

"So, my friends and I went in there, and we were just climbing around," Chris recalls. "I remember making it fairly high a couple of times. From the beginning I had no inclination to make it to the top. I just wanted to go out and have fun with everybody."

Perhaps his idea of fun deviates considerably from the mean, but there is no denying Chris' generally optimistic outlook on life, despite the tragedies in his family's past that continued right up until the week before he and his classmates undertook the Herndon Climb, when Chris went home on leave to speak at the funeral service of a close family friend who took her own life. That was the last time that his mother saw him before the day of the Climb.

Barrie was visiting the Academy to attend her son's Climb more or less on a whim, since she had barely heard of the ritual. She didn't know Kevin during his time at the Academy, and he spoke of it rarely if at all during their courtship and marriage. And with a midshipman's typical modesty, Chris had downplayed the emotional significance of the Climb, the riveting drama and the edge-of-your-seat excitement of the spectacle.

"I know it's a big deal, but I really don't know why she chose to come for it. It was kind of a random event and time period for her. But that she was there? That was really cool," Chris explains. "I think it's awesome that she got to be here for that experience, especially given the result."

"I had no idea what Herndon was," Barrie says flatly. "Chris had just been home for the funeral, where he gave a eulogy. I remember driving [to Annapolis] and I was pretty raw, emotionally."

When she walked onto the grounds of the Academy and saw the greased Herndon Monument for the first time—and found herself in the middle of a great sea of people who had turned out and were trying to find the best available viewing spots—Barrie quickly realized that Chris had been holding something back.

"After the experience I went from such an emotional low to an unbelievable high," Barrie recalls. "So, since Chris had downplayed it, saying Mom, it's no biggie, and if you've got a lot going on don't put pressure on yourself to be there, thankfully I told him no, I'm coming. I don't know what it is, but I'm coming."

Like many attendees, she was startled by the loud report of a cannon announcing the start of the Climb, and any remaining ambivalence vanished as the crowd of plebes, following the Iron Company and accompanied by a great roar of approval from the assembled happy throng of spectators, rushed excitedly to the monument.

Among the homogeneous parade of blue shorts, similarly toned and shirtless physiques, and identical Navy haircuts, Chris was easily identifiable: he was the only one of the participants sporting an enormous tattoo across his bare back. Witnessing the Herndon Climb for the first time, with her son in the midst of the sprawl, Barrie's emotions quickly ran the gamut of motherly pride, elation, and worry.

"I'll never forget it. I was by myself, and I was just, you know, *in the crowd*, having no clue at all," she recalls. "They were spraying them with hoses, and it was all greased, and . . . I couldn't believe the hoses! They kept spraying those kids. As a mom I was thinking, oh my gosh! Nail biting! Afraid they were going to break limbs! They kept falling, and tumbling." Among those falling, and tumbling was Chris, who had not thrown himself against the monument to be part of the lower layers; he was *climbing.*

After each tumble into the meat grinder, happy-go-lucky Chris began again. He wasn't a solitary climber: he had strong assistance from his bulkier classmates, who encouraged him to go higher.

"I remember I made it pretty high, because I'm a smaller guy. I probably weighed 130 pounds at that point, maybe 135," he laughs. "Guys would ask how much you weigh, and if you said a number that sounded good, they'd toss you on their back. So I said I weighed 125 just to make it sound even better."

Soon Barrie was on the phone with her old friend Vic Otero, USNA class of 1985 and a classmate of Kevin's. Barrie and Vic kept up a running commentary, the excitement of which could rival a Marv Albert sports telecast.

"I'm screaming, oh my god, Vic! You won't believe it! And he was screaming with me, and I called out, 'Chris! Put it up real high!' but then he fell down," Barrie laughs at the memory. "Vic said, 'I think Chris is going to get it!'"

Her blood pressure continued to climb as Chris climbed, and slipped and fell, again and again. "Vic said he thought Chris was going to get it; I said, I don't know about that, but it would be pretty amazing! So, as it turned out, Vic called it," Barrie recalls.

After multiple attempts and falls, Chris neared the summit for his final approach, where he found that the Dixie Cup had already been removed, or most likely knocked from its perch by a thrown, balled-up, and sodden T-shirt. His classmate Adam Connell had hoisted him up; he didn't know Adam at the time, but later they became good friends, sharing the memory of their starring and supporting roles in the Herndon drama.

"I remember just being *so* close, and [the top of the monument] was just a little bit out of my short-armed reach," Chris chuckles as he explains the moment when he held the brim of a precious combo cover in his fingertips and stretched every muscle and ligament in his body to get it as high as possible. "I just gave it a little nudge . . . and it landed."

Indeed, it was Chris who finally stood at the top of the stack of his classmates and lifted his arms triumphantly toward the heavens as a mighty roar of approval swept across the grounds, seconds before the cheers turned to the joyful chant of "Plebes no more! Plebes no more!" The official time was recorded as a highly respectable, quicker-than-average one hour and twelve minutes.

After a little random wandering in search of the best viewing spot, Barrie had made her way across Blake Road and perched on a low wall adjacent to the front steps of the Chapel, where she witnessed the final moment as Chris gently tossed the combo cover onto the very apex of the monument. She left behind her cell phone on that low wall in her exuberance to get to her son and embrace him after his triumph; fortunately, a good Samaritan maneuvered through the crowd behind her and returned it.

Between Herndon and the top of the steps of Zimmerman Bandstand, where Adm. Ted Carter stood awaiting, the crowd was particularly dense. There Chris was being carried aloft on the shoulders of his happy classmates, who were glad not only that they were no longer plebes but that the heroic Herndon scaler was a compact and wiry soccer player and not a 250-pound football lineman. Eventually the sea of humanity parted, and Barrie saw the joyful coterie bearing Chris.

"I was trying to get closer, and it's hard to get through the crowd. And I started saying 'Excuse me! Excuse me! That was my son!' and people were clearing the way for me," Barrie recalls. "And then of course it was all the excitement, and the photographers and pictures, and people were interviewing him from local TV stations . . . well, it felt like I was the mom to a quarterback who won the Super Bowl. As a parent it was one of the highlights, going down in the books."

The notoriety of being the honored Herndon scaler has always followed him, although it has fallen from its initial fever pitch, which according to Chris lasted about a month or so. He takes it all in stride, although he admits that sometimes he is confronted with a boorish broadside.

"I'm pretty sure most people in my class are aware that it was me and, to this day, I'll be in a bar or somewhere, and just to [mess around] with me someone will say, oh *that's* the guy that climbed Herndon. Like that's all I'm known for," says Chris. "People love to talk about it, but I don't think many people know my story."

His career at the United States Naval Academy is just one chapter, near the beginning of that mostly unwritten story of Chris Bianchi.

8

THE NEVER-ENDING BATTLE

(or, A History and Her-Story of Herndon)

These are the times that try men's souls. The summer soldier and the sunshine patriot will in this crisis shrink from the service of their country; but he that stands it now deserves the love and thanks of man and woman. . . . We have this consolation with us, that the harder the conflict, the more glorious the triumph.

—Thomas Paine, December 23, 1776

We must remember that one man is much the same as another, and that he is best who is trained in the severest school.

—Thucydides

MEN.

It's always about *men*.

The two passages at the head of this chapter, courtesy of the pen of Thomas Paine, British citizen turned American revolutionary rabble rouser, and from Thucydides, famously persnickety historian and general of ancient Greece, may be found in any respectable collection of great literary quotes.

They also appear on the title page of the 2018–19 edition of the Naval Academy's venerable *Reef Points: The Annual Handbook of the Brigade of Midshipmen.*

Generations of midshipmen since 1905 know *Reef Points* as more than just a handbook; it is a bible, a vast store of information and knowledge ranging from the indispensable to the obscure. Lengthy passages in the small, dense volume are not only required reading, they must be committed to memory by all first-year midshipmen, and are subject to rapid and immediate recall at the request of any upperclass midshipman.

A 1975 act of Congress authorized, or *required*, the admission of women to all service academies, including the United States Naval Academy. Since the first female graduates in the class of 1980, women have gone on to distinguished careers as Navy officers, including flag rank, as well as leaders in the business world and in local communities.

But when women enter the Academy, they are reminded, as they have been and will likely always be reminded every day of their lives, that our society remains largely and frustratingly paternalistic, dominated by men in so many ways.

Maybe things have changed since the days of dear old Thucydides, some twenty-four centuries ago, or from the times of Thomas Paine and the early American patriots, who saw women not as soldiers and political representatives, but as little more than cooks, nurses, water carriers, and flag stitchers.

But have they changed so much since 1975? Maybe not. Ask any woman of the Academy's class of 2022, who saw those two quotes the very moment she picked up her copy of *Reef Points* as plebes and began to read.

To Paine's credit, progressive thinker that he was, he managed to shoehorn the word "woman" into his quoted passage, although by comparison he uses the words "men's" and "man" and "he" three times. And the single occurrence of "woman" is only seven words apart from the word "love" in the same sentence, a common and almost automatic association of the female gender with qualities of softness and nurturing, instead of strength and soldiering. As for Thucydides, scholars in the ancient Greek language differ in their views of the modern English translation of his work that resulted in the word "man." Let's give him the benefit of the doubt and assume that he meant to refer to all people, inclusive of women.

But to a young female plebe entering the Academy, the message is subtle, and it is quite real. The reminders are always present.

She and the other women know, as they have always known: Their struggle to prove themselves just as good, just as tough, just as ready for the task of training to be outstanding Navy officers as any man will come with an extra measure of challenge. They must match the men on an unlevel playing field.

That field was never more grotesquely distorted than in the summer of 1976 when, for the first time in history, female plebes were sworn in to the United States Naval Academy.

Barbara Ives was among the first group of women to graduate from the Academy, in the class of 1980. After commissioning, she went on to a distinguished career, most notably serving a stint as executive officer on board the USS *Harkness*, a *Chauvenet*-class hydrographic survey ship engaged in a critical, year-plus research mission to the waters around Iceland, where the people are friendly and welcoming but the weather and seas are often not.

As a young officer on board the *Harkness*, Barbara was placed in command of the deck department. Which means that she was responsible for all the boatswain's mates and undesignated seamen, the most rough-and-tumble group of sailors on board ship—no disrespect meant to boatswain's mates—and not a single woman among them.

"This was back in the early '80s, and in the early '80s they said that a good percentage of boatswain's mates were told, 'Either you go to jail or join the Navy,'" Barbara says with a laugh, "so most of those guys joined the Navy."

Her Academy career left her well-prepared, despite her youth, to easily deal with such a pack of rowdies in her division on board the *Harkness*. Many, if not most of them, lacked even a basic high-school education. She didn't bother to ask for their respect; she simply commanded it. And she dedicated herself to the goal of helping each and every one of those young men obtain a high-school diploma.

"To my mind, that turned things around in that department," Barbara explains. "When the folks in the department realized that I was doing something to not just help the Navy but to help themselves as well? That was a turning point in our department."

Barbara was high in the chain of command on board the *Harkness*, and she wielded a great deal of influence. But much earlier in her career, during her first year at the Naval Academy, her position was exactly the opposite.

Those who were a part of the class of '80, male and female members alike, generally agree that they were not well-prepared for the harsh glare of the national media spotlight that accompanied their swearing-in and followed them all year long. They were inexperienced, mostly eighteen or nineteen years old, fresh out of high school. They had met the Academy's rigorous entrance requirements, so they were a cut above average, but they faced a demanding program that was designed to test the boundaries of what these "best of the best" even thought was possible. The dropout rate during plebe year is always sizable; successfully finishing the year, constantly beset by challenges to body and mind, is difficult enough for the hardiest among them without the added pressure that was unique to Barbara's class, the first to be gender-integrated in the Academy's history. With no women yet among the upper classes, those trailblazing female plebes often felt as if they were isolated on an extremely lonely island.

As the academic year wore on and Commissioning Week approached, Barbara and all her classmates were well toughened, having proven themselves worthy of continuing their Academy training. The women among them had grown accustomed to the whistles and catcalls, the unsubtle jokes and taunts, the whispers and stares and silent looks that they had expected since Induction Day and that had been delivered in ample quantity throughout the year.

Everyone knew about their upcoming date with Herndon, and what they were expected to accomplish in the Climb. Still, anticipation of the event brought about a new and unanticipated wrinkle.

With the long-standing policy that plebes leave their room doors open during the day still firmly in place, conversations were easily overheard, in some cases perhaps purposely so.

"There were a lot of discussions in Bancroft Hall. Not just upperclassmen talking among themselves," Barbara recalls, "and you could hear them in the corridor. There was a lot of conversation about Herndon and [the idea of] no girls on Herndon."

That phrase would shortly become ubiquitous, and it soon became the notorious acronym "NGOH," a code that was a secret to no one.

"They didn't really say NGOH just yet, they were just saying no girls on Herndon," Barbara explains. "A week or so before Herndon you could hear those conversations going on. And, of course you were always teased by a few of the upperclassmen, with 'Hey, you're not going to make it to Herndon! We're going to make *sure* you don't get to Herndon!'"

She insists that Naval Academy leadership had no role in this, but the sentiment was nevertheless widespread among midshipmen, both plebes and upperclassmen. In fact, leadership was adamantly and publicly opposed to such discriminatory talk.

"After Herndon, I learned that the commandant had put out word to all the company officers that if they had heard about 'no girls on Herndon,' or saw anything 'no girls on Herndon' they were supposed to report it right to the commandant," Barbara continues. "So I was aware of that, and I was also aware of a few upperclassmen that were pulled into the commandant's office because they were caught selling T-shirts that said No Girls On Herndon, or NGOH." These shirts were never worn publicly on campus, but many are still held in private collections.

To the women, the existence of such T-shirts was an obnoxious reminder of the treatment they faced on a daily basis, but it would hardly affect their determination to fully participate in the Climb.

Barbara Ives was one of the most active and aggressive climbers. "I just wanted to be a part of it. It just felt like it was my responsibility to do what I could to help out," she says, "supporting my classmates, moving forward and climbing that monument. It was like a tower. . . . I could actually call it a tower, because once you got started you looked up and said, oh my goodness, now it's more like a tower. That's what I remember."

She was forewarned, in both friendly and unfriendly terms, and in those overheard corridor conversations, that women who attempted to participate much beyond a spectator role would not be helped up or pushed up. Instead, they were likely to be pulled down.

Tom Lindblad, another member of the class of 1980, saw that women were being actively, physically discouraged from reaching the upper levels of the human pyramid around Herndon. And to his classmates, the ordeal was sufficiently challenging without an added measure of sabotage from within.

"It's probably one of the hottest times I've ever experienced outdoors," recalls Tom, who retired from the Marine Corps with the rank of lieutenant colonel after serving twenty-two years, including a deployment in support of Operation Desert Shield. "From everyone's body heat, sweating, [humidity]. And you've got about a thousand people shoved inside an area maybe about half the size of a basketball court." Like anyone else, he was mostly eager to see the Climb come to an end. And when a female climber was pulled down, there was often collateral damage, as other nearby bodies lost their footing and tumbled down. Estimates differ as to how many of the men of '80 were actively engaged in such counterproductive measures, but just a handful of strategically placed saboteurs continually disrupted the progress of the building of the structure of plebes that would eventually reach Herndon's apex.

"We were there for three and a half hours," Tom says, bristling slightly at the memory, "and there's only so long you can be at the base, with people on your shoulders. It's good team building, a cooperative experience, and that's the kind of feedback you want from it, but if [many of] your thousand people are working against it, it's chaos."

During her plebe year, Barbara Ives fought her way through many obstacle courses. The task of scaling Herndon was just one in a series of many, she imagined as she scampered upward on the backs and shoulders of willing classmates.

And at Herndon, she found a winning strategy.

"I started going in a zigzag motion, up the ladder, actually up the human ladder," she explains. "I was climbing up. I was going in this zigzag motion just so when someone reached to pull me off, they couldn't; I had already moved to the left or the right, so they just sort of grasped my shirt, or just grasped air. So it seemed to work pretty well."

The agile and elusive Barbara spider-walked up the pile, falling down a few times, either yanked by an ankle or by simply losing her balance and footing. Undeterred, she was right back at the Climb after each unsuccessful effort.

"Before I knew it, I was at the top. I didn't even realize . . . I'd gotten to the top and I was touching the Dixie Cup," she explains. "I thought, 'oh my gosh, I made it!' Before then I was just trying to keep climbing and not get pulled down."

In a typical bit of upper-class sadism, the sailor's cap was not easily removed; it had been glued to the top of the monument with a high-strength bonding agent. A small bit of cloth tore off in her hand. She reached for another, and as someone below tried to pull her down, she held on for dear life, clinging to a larger, loose piece of the glued-down Dixie Cup. A male classmate had hold of it too, and a struggle for possession ensued, as if the simple sailor's cap were the queen of England's royal crown.

"It was really well attached," Barbara recalls. "I was in a kind of tug-of-war at that point, against who was pulling me down, and holding on to that piece of the Dixie Cup. Eventually, the cap came loose, it came off, and I just fell backwards and fell off the monument."

That was her last chance. A few moments later, the monument was capped with an officer's combination cover and the plebe year, for both men and women, was over. Barbara would have to be content with the knowledge that all she did was make history: in the first Herndon Climb with women participating, she made it all the way to the top and touched the well-secured Dixie Cup with her bare hands.

"I think that had I not been pulled off, I would have probably shared the honor [of removing the Dixie Cup and putting up the combo cover] with my male counterforce classmate," Barbara laughs, "but just having participated in removing the cap was a wonderful accomplishment."

And Barbara shared in all the joy of her classmates. "A short time after I fell down [the combo cover was placed] and we could say that Herndon was done, and we could move on."

They could move on with a new status and fresh enthusiasm for the challenges ahead, no longer plebes, no longer the lowest of the low.

By the time Pam Pitkin, class of 1982, took the oath two years later and entered the United States Naval Academy, two gender-integrated classes had already successfully advanced, although the dropout rate was significant. The presence of increasing numbers of women on the Yard may have been some comfort to Pam and her female classmates, but it did little to alleviate the constant demands of plebe year.

Pam was assigned to 27th Company, which had a reputation, well-earned in her opinion, as a very tough, brutal company.

"The way I've gotten through life is not necessarily on my brilliance or charm," she laughs, "but on sheer guts, determination, loyalty, a thick skull and a tough hide."

Pam suffered a traumatic head injury on a high school softball field, leaving her with difficulties in memorizing things by reading, seemingly a critical skill for success at the Academy and in the Navy. She is candid and frank in discussing the difficulties she faced and the ways she adapted to the demands of the academic year among the 27th, from whom she asked no quarter and received none.

Any memorization exercises, whether reciting sections from *Reef Points* or the evening's dinner menu, or memorizing important points in classroom lessons, were an extra challenge for Pam.

"I could remember things, but just not the way I needed to or how the Academy needed me to," Pam explains, recalling some of the frustration she felt during plebe year as demerits mounted for minor infractions. None of those, however, were given for flaws in her uniform, as she maintained focus on her personal strengths, rather than weaknesses.

"The only thing they couldn't take away from me was essentially my ability to look outstanding. Polished brass, polished shoes, outstanding uniform appearance," she says. "I could look good, but I just couldn't memorize things. I learned by seeing and watching. I had a really tough time learning from books."

With demerits came punishment, meted out largely in the form of weekend "area tours," endless hours of marching back and forth in an obscure and drab open courtyard. Her parents lived only an hour's drive from the Academy, yet Pam was seldom able to get home, spending almost every weekend of plebe year glumly and morosely, on area tours.

Pressure to drop out of the Academy assailed her every day, starting on Induction Day, including from classmates in her own company.

"I found out that five of them approached my roommate and asked if she was going to help get me out of the Academy. This is on Induction Day! Eventually, none of those people graduated, including my roommate. And I did!" Pam says with no small measure of pride.

But before she could graduate, she had to get through four years, and the first year was unrelentingly demanding for Pam, with each day bringing a parade of upperclassmen screaming at her, pushing her limits, going beyond the norms of decency. "I had a year. A lot worse than many of my classmates, than all my female classmates. But I was there, because that's who I was. I truly believed in duty, honor, country, and I was 100 percent just who everyone thought I was. . . . I was serious, never learned how to play or goof around. . . . I never compromised my standards. And because they were bound and determined to get me out, I was bound and determined not to let them."

By the time the long year neared its end, Pam's grades were more than satisfactory to permit advancement, and the tide of demerits had slowed to a trickle that would be too small to cause her expulsion if it continued at its current rate. But Herndon still loomed.

"That Climb meant more to me than it did to a lot of really smart people, a lot of people who didn't have it very hard," Pam says with a touch of wry resignation. "There were some people that went [to Herndon] and it just wasn't a big deal to them. There was no clear line of demarcation between those who had brutal hell for a year and those who were . . . *not* in that circumstance."

And although two years had elapsed since the women of the class of '80 had been the first to participate in the Climb, the seeds of the notorious No Girls On Herndon movement were still alive, spreading their ideology, secretive in the corridors of Bancroft but in full bloom at Herndon.

Pam and her classmates were well-prepared for what they would experience, which was a distillation of all the resentment they had felt throughout the academic year. "We took a lot of physical hammering, intentional hitting, elbows, being jerked around by the collar, being pulled down, trampled, that kind of stuff," Pam recalls, "but I was big enough, and I did hit back. If you want to play that game, that's fine. But quit picking on the little ones."

The challenge was deeply personal for the women, and never were gender lines brought into sharper focus than during the Climb.

"The entire atmosphere was different between the men and women," Pam says, "and there were good men out there just like there are good and bad among everybody and everything. But seeing my classmates pulled down and dragged, getting thrown off the monument as they were trying to get up, was pretty disappointing. It wasn't a surprise. And it was just the beginning of what we were going to get."

One of the Climb's more popular, recent fashion customs took on a sinister shade from the perspective of Pam and her classmates. Long gone are the days when midshipmen—all *men*—wore their daily uniforms to swarm around Herndon in the "snake dance" after the posting of final grades and when they donned work whites to climb an ungreased monument. Nowadays, after the plebes assemble in formation in their blue athletic shorts and blue rim T-shirts, most of the men immodestly strip off those shirts, showing the results of so many hours of calisthenics and endurance training. Many of the women, too, are proud to show off their flat stomachs and well-chiseled arms and shoulders; the shirts are ostensibly used to remove grease from the monument to help ease the Climb, but there is a subtle and ever-present tonal difference in such showing off, and it wasn't lost on Pam or her female classmates coping with the lingering No Girls On Herndon sentiment.

"It's a totally different thing for men and women," she explains, observing that the men pulling off their shirts was "much more like the precursor to a sexual assault, as opposed to 'ha ha, let us have your shirt to help do something good for the class.'"

A post-Herndon photograph shows Pam, covered in mud and grease, flashing a million-megawatt smile. The pride of getting through a brutal plebe year, and of graduating from the Academy three years later and serving as a Navy officer, are feelings that she will always carry with her. But the pride is mixed with bittersweet memories for Pam and for generations of women who preceded her and those who followed her.

One particular, prominent individual contributed as much as anyone to the gender gap and to the discord and distrust that plagued the Brigade of Midshipmen during the difficult period of adjustment in the late 1970s and early 1980s as the Academy struggled to seamlessly incorporate women into its ranks.

James Henry Webb Jr. graduated from the Academy with the class of 1968 and went on to serve in Vietnam as a Marine Corps Infantry officer. His heroism earned him the Navy Cross, the Silver Star, two Bronze Stars, and two Purple Hearts. After leaving the Corps with the rank of captain, he attended law school at Georgetown University, earning a Juris Doctor degree and a Horan Award for excellence in legal writing. A prolific and gifted writer, he is the author of ten books, both fiction and nonfiction, and he has even contributed to the screenplays of several major Hollywood films. In 1984, President Ronald Reagan appointed him to the newly created post of assistant secretary of defense for reserve affairs, and in 1987 he became the first Naval Academy graduate to serve in the civilian position of Secretary of the Navy. His political career further blossomed in 2006 when he was elected United States senator from the state of Virginia, and in 2015 he made a tentative bid for the Democratic Party's nomination for president.

In 1979, Jim Webb penned an extraordinarily divisive opinion piece for *Washingtonian* magazine titled "Women Can't Fight," following it with a major speech on the same topic before the entire Brigade of Midshipmen, which became a watershed moment for women at the Academy.

As a sophomore, Catherine Ripley was there to hear that speech, and she will never forget it, particularly a certain moment of audience participation that hit with the force of an artillery shell.

"There were thousands of midshipmen in the room, and [Webb] was talking about female midshipmen graduating and becoming doctors in the Navy, Navy doctors. And Webb said, well, we're not training you to become doctors. We're training combat officers."

Then a male upperclassman raised his hand and asked, "So why are women at the Naval Academy?" And according to Catherine, nearly the entire brigade stood up, en masse, and broke into applause, with the exception of all the women and a handful of men.

She was seated next to Jay Gallo, the Herndon "scaler" for her class and a teammate on the sailing team. "We were very close friends," Catherine says, "and I just kind of glared at him, and he did *not* stand up and clap. There were [some] that didn't, men and women, but most of them stood, and the applause was thunderous, and it cut. It cut deep."

Catherine's recollections of her experience at the Academy differ from those of her 1982 classmates only in the names and faces and details. She bristles at the mention of the despised NGOH acronym, and at an even more notorious and disreputable one: WUBA, meaning "Woman Used By All."

The two acronyms entered the lexicon and the Academy concurrently with the admission of women, but the latter one has a slightly more tortured origin.

"I don't know who invented it, or what class, but it was used ubiquitously throughout the Naval Academy when I was there," Catherine explains. "This term WUBA, was 'Working Uniform Blue Alpha,' which was just the uniform that we wore to classes every day. And they were actually black, I don't know why they called them blue, but anyway, a WUBA was a Working Uniform Blue Alpha, but they changed the acronym, and used it to refer to women, *all* women at the Academy. It was certain men; they didn't all call us WUBAs, but they changed the acronym to 'Woman Used By All.'"

Even after a twenty-six-year career in the Navy, retiring with the rank of commander and an extensive record of service in the diplomatic corps and as a foreign area officer, some wounds still sting, such as the ones administered by fellow classmates, who were honor-bound to treat everyone respectfully but often did the opposite.

"They know it hits. It's kind of like a cut. It's death by a thousand cuts, what we went through at the Naval Academy with things like that," Catherine says frankly. "With references to women, [that we shouldn't] be there. I was told, 'You should go home, back to Montana and be a cheerleader.' And a lot of other weird things."

As a plebe, Catherine knew some of the women in the two classes ahead of her. But the seniors, members of the class of '79, were entirely male, and they were well aware of their role in history as the last Academy class to be an exclusive men's club. They took a certain pride in that distinction, going so far as to adopt the Latin phrase *omnes viri*—which translates to modern English as "all male"—as their slogan and having it engraved on their class ring.

"They didn't hesitate to rub that in our faces. Literally, they would shove it in the women's faces," Catherine recalls. "The firsties were all male and because there were very few women officers that were on the staff, and there was still a lot of unhappiness that Congress passed legislation to allow us in. Most of us would roll our eyes. At the time, we didn't know what was acceptable and what wasn't, so we just accepted all of it as harassment. I think people did not come to terms with the reality that we were there to stay."

The casual and careless harassment would continue through plebe year and the rest of her time at the Academy and her naval career and into civilian life. But it came into particularly sharp focus as the class of '82 was confronted with the challenge of the Herndon Climb.

"The buildup to Herndon was huge," says Catherine. "There was a lot of excitement, and a lot of nervousness, and a lot of anticipation. None of us thought anyone would get hurt, and it was just excitement! When we get [to the top of the monument] it's a symbolic coming together as a class, doing something as a team, and to beat the previous classes, right? But we didn't work as a team."

In this aspect, Catherine's plebe class was no different than the two that came before hers, and many of those that came later: the infamous No Girls On Herndon movement reared its head, time and again, resulting in women being discouraged from participating in the Climb and in being pulled down as they attempted to do so, often to the detriment of the goal. But putting aside the discrimination and personal resentment, Catherine takes an analytical view of the practical aspects of the Climb.

"If you're really going to be engineers and figure out the problem, you would want the lightest people to get up there. So a five-foot-four guy is going to be about a buck twenty, or whatever," she laughs. "We got to the top, and [Jay Gallo] is pretty short, he's five four, and we all knew that because it was the length of the keel of one of our boats!"

Unfortunately, too many male classmates saw the Climb in different terms. Considerations of logic and engineering principles were cast aside, and instead they focused on maintaining male dominance throughout the culture of the Academy; and for them, the Climb presented an opportunity for women—hardened, strong, and highly competitive women—to threaten their notions of supremacy.

"There were many girls, like who were like five-two and weighed barely a buck, or maybe like 105 pounds, and we were scrappy, and we were all athletic," Catherine continues. "I mean, we got into the Academy easily, so we had some form of athleticism anyway. And there were a lot of women that did get halfway up, and they were literally pulled down, yanked down and thrown to the ground. We were boosting up the scrappiest and smallest of us, and I wasn't one of the smaller ones because I remember pushing a couple of my female classmates up, and we thought everyone should push them up . . . and they would get thrown down and we would push them up again. It became kind of ridiculous. We were covered in grease and mud, and eventually we just gave up, thinking someone's going to get hurt, and we can't keep doing this. I have vivid memories of women that were [pulled down] and one of them was really tiny. She was one that got really high up and she just got thrown off, her little tiny body."

And why didn't anyone in a position of authority put a stop to it? Why didn't anyone advocate for them? Catherine and many women wonder why not.

"No one said anything. Everyone saw it happening, there were several company officers, all majors and lieutenants, lieutenant commanders were out there. Commanders, battalion commanders. *Nobody* stepped in," Catherine laments. "At the time there were very few, maybe two or three female officers on the Naval Academy staff. A few teachers and one company officer, I believe. We didn't have any kind of mentorship or leadership to step in or to say anything. And this was the third year it had happened."

They were far outnumbered, and they couldn't look to the chain of command to help them, but true to the overriding mentality of good Navy officers, the female plebes of Catherine's and other classes made certain careful preparations to guard against an even worse potential outcome, which would have been much more humiliating than being thrown to the ground.

"[Before Herndon] the women all got together and said this is gonna be crazy and we're gonna have fun, *but* we had the wherewithal to band together and agree that we all needed to wear our issued one-piece Speedo bathing suits under our shirts and shorts. We thought our shorts and our T-shirts might be pulled off, and we could be left bare, hanging off of this greasy pole . . . so we passed the word, 'you don't want to be caught in your underwear. Or worse!'" Catherine laughs.

But laughter often hides an undercurrent of anger and resentment.

Karen Halverson recalls the warnings that she and her classmates received before her class of '84 undertook the Climb.

"Our firsties [seniors] were class of '81, and most of them were really great—professional yet caring—and they really tried to help us out as plebes," says Karen, whose last name was Malloy during her Academy days. "I distinctly remember them telling the women, under no circumstances were we to get near the Herndon obelisk during the Climb. They told us that women in previous classes who had climbed up the human pyramid around the monument had been digitally penetrated by the men beneath them as they stood with their legs apart on the shoulders of those in the tier below."

Whether or not the claim was factual, such bone-chilling words of warning achieved their desired effect on Karen and many of her female classmates.

"I never met anyone who would admit that this had happened to them, but I remember listening to this horror as a very naïve nineteen-year-old, and it was almost beyond comprehension for me," Karen recalls. "Needless to say, my roommate and I went nowhere near Herndon. We watched from the perimeter and our only contribution was to throw the shirts, that we wore over our bathing suits, in to those who would use them to wipe off the lard."

The men participating in the 1981 Climb, plebes of the class of '84, held a variety of opinions concerning the integration of women into the Academy, as ordered by the United States Congress, and the memory was still fresh in Navy circles, rankling some while others took it more or less in stride. As a consequence, the women were ignored by some of the men, helped by others, and loudly scolded by still others.

"I remember male classmates yelling at us because we obviously were not participating, which probably did not do much to improve the negative feelings they most likely had about women being at the Academy!" Karen observes. "It was an interesting sense of complete detachment, and revulsion, to a tradition that had so much meaning for the vast majority of my class."

Karen served for ten years, achieving the rank of lieutenant commander before resigning her commission. Her distinguished career includes a post as command center officer for the Pacific Fleet, in charge of deployment of communication equipment and satellite networks for the eastern Pacific.

Nowadays, she devotes her time to various philanthropic organizations dedicated to serving military veterans and their families and serves as a mentor for young military spouses.

"Something I always find remarkable," says Karen, "is that I never encountered the hatred, meanness, discrimination, and attacks in the Fleet that I did at [the Academy], from mids and faculty alike. I always thought it was amazing that the sailors and chiefs in the 'real' Navy had so much more sense and brains than some of the jerks at school. While my memories of the Naval Academy are not fond, my memories of my time in the service are terrific."

From the suffragettes of the early twentieth century, who fought for and ultimately won the right for women to participate in elections of their government representatives; to the first female members of the Navy, who filled critical noncombat jobs in both world wars; to pioneers like Janie Mines, class of '80, who as the first black female graduate of the Academy battled against a double dose of discrimination; to more modern times, the struggles and triumphs of women on an unfairly angled playing field continue.

By the time Nikki Battaglia and her classmates of '96 took the oath and joined the Brigade of Midshipmen, nearly an entire generation of women had attended the Academy. All that time and experience brought a great deal of change to the regulations and unwritten customs governing the status of the commingling of genders in the classrooms and corridors and on the training grounds.

Regrettably, much remained the same, and a great many men continued to stubbornly oppose the very notion of female midshipmen. More subtle forms of discrimination were still mostly the norm. And many of the changes in procedures and regulations turned out to be for the worse.

In the days leading up to Herndon, the women still got together to address concerns—and warnings—about the nature of the ritual. Only for Nikki and her classmates, the get-together was not secretive or surreptitious. Now it was formalized and officially sanctioned, leaving the underlying message largely unchanged, which was perhaps even more discouraging to the ambitions of female plebes who eagerly sought to fully participate in the Climb.

In accordance with long-standing tradition, the male plebes were given only the most cursory and unspecific briefings from upperclassmen, amounting to little more than a brief summary of the rules, along with a modicum of fairly useless gossip. Perhaps unsurprisingly, the women were treated differently.

In the days when the Navy was still reeling from the infamous Tailhook scandal and its aftermath, when the widespread harassment of women throughout the ranks was splashed across the front pages of daily American newspapers, female midshipmen were accustomed to briefings that were separate from those of their male counterparts. According to Nikki, however, this pre-Herndon briefing took an unexpected and frustrating twist.

"All the women were pulled aside and talked to separately. I remember when they called all of us together, a group of about 120 of us. It was an all-women brief, just things that were specific to us," Nikki explains. "It was a bunch of upperclassmen and some of the faculty. They told us we had to wear our bathing suits under our clothes, and they told us we had to wear underwear, which I thought was quite amusing. And we had to wear our hair up."

Nikki goes on to describe a meeting that she recalls in vivid detail after nearly twenty-three years of service as a Navy officer, with seventeen years flying helicopters and more than five years as an aerospace engineering duty officer with the Naval Air Systems Command. The meeting had all the trappings and appearance of a typical "pep talk," but to Nikki and her classmates, the advice seemed to be that they should largely avoid getting involved in the dirty and greasy work of Herndon and take on a role that would amount to little more than that of cheerleaders at a varsity football game.

"They told us that we would get pulled on, and so we had to wear a bathing suit to make sure nothing nefarious happened, I guess," Nikki says, "and basically [left us] with the impression that they were almost dissuading any of us from participating, because they didn't want anyone's shirt to come off or any girls to be groped or anything like that, so they were almost saying we should just sit around on the sidelines and watch the guys do it."

By now, the No Girls On Herndon sentiment was widely known and widely condemned in official circles. According to Nikki, all the plebes, male and female, were told that they could write or draw on the blue rim T-shirts they would wear for the Climb, but that permission came with a strict admonishment.

"We couldn't have any bad words or anything degrading to anybody on our shirts," Nikki explains. "I had heard some of the talk about 'no women' or something like that, but they didn't say anything [specifically] about that. They just told us, if you're going to write on your shirt, it's got to be something supporting company unity." But it was clear to all that any sentiments that targeted women were forbidden and violators would be punished, possibly charged with a conduct violation or dismissal from the Academy.

But after the plebes of the class of '96 assembled in Tecumseh Court by their companies and rushed the monument after the sound of the first cannon blast, it was business as usual at Herndon, no different from the situation faced by the earlier pioneers like Barbara Ives, Catherine Ripley and Pam Pitkin. And just like those women, Nikki and many of her classmates were determined to prove themselves, and they simply refused to abide by the repeated warnings and discouraging advice they had been given.

"I started climbing up on people, and getting up there, maybe two or three people high, and I got pulled down. And you know what? I climbed up again," Nikki recalls with a healthy dose of pride. "A lot of girls in my class thought, let's just try it, and a lot of them climbed up, but we all got pulled down. A few of us made it three or maybe four levels high but we all got pulled down. I think we all knew that a woman wasn't going to get to the top, but we thought we'd give it the old try, anyway."

Still, despite the pre-Herndon words of discouragement from faculty and upperclassmen, Nikki was surprised by the determined opposition to her efforts and to the shameless disregard of many of the men's efforts to work together effectively, as an entire class with their indispensable female allies, to accomplish a critical objective. All the principles of teamwork, loyalty, and honor that had been hard learned at the Academy suddenly seemed, to the women at Herndon, to be nothing more than hollow sentiments.

"I don't think we expected to be *pulled* down," Nikki explains, "because many of the guys would climb up and fall down, because we're all greasy and it's natural that we're going to slip and slide. But I don't think I expected to actually have someone reach up and pull me down. I remember being a little ticked off, but it's been twenty . . . something years," she says with a laugh, "so I'm not *as* disgruntled about it now. We all knew they weren't going to let a [woman get to the top], and to this day, there hasn't been a woman up to the top of Herndon."

Knowing that their efforts were ultimately doomed to failure, Nikki and her female classmates nevertheless did their best to establish a new benchmark for women's success in the Climb, a proverbial giant shoulder for future classes to stand upon.

With coordinated and determined effort, the female plebes tried to demonstrate the potential effectiveness of a strategy to form an entire ring around the monument, layers above the established levels of men in the lower reaches of the human scaffold, that would hopefully endure long enough for one of their own to reach the apex, and successfully place the combo cover.

"There was a ring all the way around, basically one complete level, so if there had been somebody above that level, they would have been able to do it. And it was all women," Nikki says. "And just like any Herndon Climb, the rings last for about seven seconds, and then everybody comes down."

Seven seconds. An unscientific measurement for sure; but if a group of women were able to achieve that record and surpass it by one or two or more seconds, with just a little assistance from the level below and a little less interference from the men who were determined to oppose any woman who sought to represent her class and enter the history books as the first female Herndon scaler, it will happen, as surely as the Severn River flows past the Academy.

Shannon Murray has a uniquely well-informed perspective on the changes in Academy culture and procedures that developed over the course of a couple of decades, and she saw how those changes became manifest at Herndon.

When she graduated from the Academy with the class of '92, she was Shannon Lacy; soon after commissioning she married Scott Murray, himself a USNA grad from the class of 1990. Not unexpectedly, both spouses are well aware of their respective classes' official times in scaling the Herndon Monument. For those keeping score, like the Murrays certainly do, Scott's class was faster than Shannon's, by all of ten seconds.

And Shannon had a typical experience at Herndon. "During my plebe Herndon Climb, I saw classmates pull women off the pile. I didn't even really try to get on the pile, and most of my female classmates stood back with me. We didn't want to get hurt," she recalls. "Only 10 percent of our class were women. And they were telling us, don't even try, because you're going to be pulled down."

This came as something of a cultural shock for Shannon, who was never instilled with a tendency to shrink from a challenge.

"My parents were schoolteachers; my dad was a coach and I played some sports. And I was never told I couldn't do something because I was a girl," she explains. "So when I got to this place called the Naval Academy, and I was told essentially that I couldn't do [the Climb] . . . *because I was a girl* . . . I thought, what the [expletive] are you talking about?"

Shannon and Scott now have two sons, the older of whom graduated from West Point, making him something of a "black sheep" in his Navy family and significantly increasing the stakes and the level of tension around the Murray household during the annual Army-Navy football game. Their younger son followed more closely in his parents' footsteps, entering the Naval Academy with the class of 2020. So Shannon was present, and she watched her son's class undertake the Climb, with that special level of interest that is unique to a parent.

"I saw a lot of strong young women climbing up, and even being helped by classmates," Shannon recalls. "I didn't see one being pulled down. They were using all of the strengths in their class, not just the male ones. I saw girls, women, up on the top and people were helping them get there. One girl, she was strong, I mean with muscles that were just bulging out."

Clearly, from Shannon's perspective, the sensibility of carefully engineering a proper structure—with the heaviest and strongest building blocks at the bottom and the lightest, most agile climbers unleashed around the perimeter and *not* yanked down from below—came into practice.

"They [the class of 2020] realized that some of the 'cheerleaders' and the really tiny people could get up higher," she says. "That was kind of cool. They didn't do it as fast as us, but you know . . ."

Given the opportunity to publicly boast that both her and her husband's classes posted quicker times in the Climb than their son's, Shannon declines. But privately, around the Murray household, the subject is never far out of mind.

Meagan Flannigan can testify to the dichotomy that continues to define the culture of gender relations at the Academy. As a member of the class of 2003, she saw both extremes of male attitudes laid bare at Herndon.

"My Herndon experience included some of the best and worst from my classmates," she says. "After the first fifteen minutes of people trying to build the second tier, I had been unable to climb, to join my classmates, until a friend of mine who was a football player saw me getting pulled down, and he said, 'I'll get you up!' He repeatedly helped to make sure I was not knocked or pulled down, so I could climb up to the second tier."

But for Meagan, the worst was yet to come.

"For the most part, I was treated like one of the guys, all of us just scrambling and working together to get a tower erected," she recalls. "However, shoes were being thrown at me. . . . It happened frequently enough that it was not a coincidence. Shoes were being thrown at me. One of my most vivid memories was seeing a guy who was my chemistry lab partner, always a really quiet, nice guy in our class. He looked at me when I was on the second tier and screamed 'get the [expletive] down!' I will never forget how his true colors and bias came out in that one statement."

As a plebe, Meagan was well-accustomed to hearing men scream at her, so she was unfazed by this singularly unpleasant encounter.

"I was not deterred. I was an active member of the team," Meagan states flatly. "At one point the mid who was going for the cover stood on my head to try and grab the Dixie Cup, and eventually I handed the combo cover up to the mid who was on the third tier. Overall, I have nothing but fond memories of the experience and what it taught me about resilience and teamwork."

With baby steps, two forward and one back, sometimes stumbling, sometimes falling, the culture of the Academy, and in American society in general, moves forward.

And as always, the never-ending battle continues.

9

SECOND TIME'S THE CHARM

(or, Herndon Plays Ball)

If you're going through Hell, keep going!
—attributed to Winston Churchill

He was flat on his back.

Exhausted, dehydrated, mentally drained, his long limbs trembling from being pushed to their limits of strength and flexibility, the Navy basketball team's star center heard shouts and cheers that coalesced into a noisy din. He closed his eyes for a moment, his consciousness in some doubt, as the pain in his ankle overwhelmed his sense of urgency in determining whether it was broken or merely sprained.

Basketball was the farthest thing from his mind at the moment. All he wanted was to close his eyes and rest for a few seconds. The game was over, or so he thought. If not, then he supposed he could rely on his teammates, just as he had been trained to do, not only in endless hours of basketball practices, but throughout the entirety of the rigors of plebe year.

"Just a few seconds," Joshua Caleb Williams thought, or wished, as his eyelids fluttered, suggesting that a full six hours of sleep would have been preferable. But even those few, precious seconds would be denied.

"Josh! Josh! Are you okay?!" someone was yelling.

"Get up, Josh!" someone else implored. "We gotta get you up there again!"

Josh felt that he had done his part. He made his contribution to the team's victory, and now he was not only exhausted but injured, perhaps seriously. In his estimation he should do nothing more than rest for a moment and then seek medical attention for his painfully swollen ankle. He needed rest, maybe a swig or two of Gatorade, and more rest.

But this wasn't a basketball game, and Josh wasn't on a basketball court. He was on a patch of grass, which was rapidly yielding to mud, and the mud was at the foot of a great, stone monolith with raised lettering spelling out the name of "HERNDON."

And it wasn't a hard body-check from an opposing team's power forward that put him on the ground. It was the Climb.

"Come on, Josh! Get up!" his classmates continued to urge him.

The so-called rules of the Herndon Climb are imperfectly and sometimes inconsistently enforced. Over the years, subtle interpretations had occasionally been necessary, with the superintendent acting as the ultimate arbiter, and his decision was final.

On a handful of special occasions, an on-the-spot rule interpretation has been made by the plebes themselves. In each case, when two courses of action were presented themselves and both were apparently within the "rules," inevitably the course chosen was the more difficult one. Like the mighty Hercules in Greek mythology, who, when offered a choice between a life of ease and decadence or one of labor and troubles, chose the latter, midshipmen tend not to shrink from a challenge, but rather to embrace it, and the class of '99 was resolute.

At issue was the following question: If someone removes the Dixie Cup but then falls down, can another climber place the combo cover? Or does the honor of placing the cover belong to the person who got the cap?

Josh had been the first to reach the apex of the monument and snatch the sailor's cap, scampering up a human ladder, of which the square set of shoulders belonging to his basketball teammate Seth Schuknecht made a sturdy if slippery rung; below Seth was the battleship-class body of football lineman Derek Peterson. One or two combo covers were being lifted up, passed hand to hand, rising layer by layer in his wake. But before Josh could grab hold of one of them and place it atop the monument, something slipped, somewhere down among the supporting beams below Derek. The structure fell apart, and Josh was plunged into the maw of a collapsing pile of bodies.

A couple of decades removed from his Academy days, Josh recalls the Herndon Climb with a measure of pride, a touch of fondness, and a great sense of relief that the experience is well behind him.

"At over two hours in, I had pulled the cap down, and before I could replace the cover, the pile collapsed, forcing us to start over," Josh remembers. "My body was spent, and with what I had left, I worked my way out of the mob and laid down."

His legs were fatigued to the point of near-paralysis, with the only upside being that a certain degree of numbness replaced what should have been a much more sharp and stabbing pain in his sprained ankle. Dehydration worsened the exhaustion, and the possibility of a total shutdown appeared on the horizon.

"I can't remember ever being more worthless, physically," he recalls, "but as I lay there a few of my classmates came to me and said that since I had removed the Dixie cover, I needed to be the one to replace it."

And so the question had been decided. The verdict was in: *one* person, and only one person, must top the monument, and that person must be the one who affected the removal of the Dixie Cup. Any talk of sprained ankles and dangerously plummeting serum electrolyte levels sounded like excuses, and everyone knows that a midshipman doesn't make excuses.

But there was *one* person, and only one person, who dissented from the majority opinion. He disagreed with the decision, and he was on the verge of pointing out that it was based on a flimsy legal premise, lacking precedent. But the mob immediately and loudly shouted him down, and he was not allowed to state his case.

That person was Josh Williams.

Josh had a better-than-average understanding of the limitations of the human body, particularly his own body, which had just withstood an uncompromising plebe year and its attendant physical demands that are meant to identify and expose those precise limits. He could barely sit up, much less be expected to climb, for a second time, the rickety scaffold of classmates that was hastily being rebuilt.

Those classmates were a force to be reckoned with; to borrow a well-known phrase, they had not yet begun to fight. And despite his presumption that he could declare himself out of action, the struggle against the great granite opponent was about to become deeply personal for Josh Williams.

A new battle plan was quickly developed and implemented. "Somehow amidst the chaos, this message spread, and my classmates started to shout in unison. Something had changed. There was a new energy among the mob," Josh explains.

Perhaps he was driven beyond the farthest frontier of endurance by that ever-present sense of honor, a devotion to strive, to achieve more than anyone ever expected of him, to prove his excellence as a midshipman. Perhaps his classmates carried him the remaining distance after that deep reservoir of strength, driven by honor and courage, finally ran bone-dry.

"I was picked up and forced to the monument," Josh admits, downplaying his own agonizing contribution to the group effort. "Next thing I know, I am being literally pushed up Herndon with the cap in my hand. Before I realized what happened, I was only a few feet from the top."

Those few feet may as well have been a mile. As Josh stretched his long, sore, painfully cramping arms and legs, he still couldn't quite reach the apex of the monument. Herndon, it seemed, still had plenty of fight left in him, and Josh felt that he was in grave danger of losing the battle.

For a man who stands six feet eleven inches in his stocking feet, being unable to reach something is a peculiar feeling, and it's one that Josh will never forget. He could easily slam dunk a basketball in a regulation-height goal, ten feet high, if he was in the right position, even with opponents surrounding and battering him, but getting a combination cover to the top of a twenty-one-foot hunk of unforgiving granite was proving to be a vexing problem that bore little similarity to the challenges of the hardwood court. Somehow, he would need to toss the hat, and hang it on the apex, just a temptingly short distance away, before disaster intervened. And just as surely as the sun rises in the east, that disaster would inevitably occur before Josh got a second shot.

"I sized up the distance, and I quickly realized that I had no balance to be able to make a good throw. In the next moment, I could feel the pile start to collapse again," he explains. "I still remember everything going silent in my mind, and it suddenly felt like slow motion as I took the shot."

With the Navy team on the basketball court, Josh was a tenacious defender, an intimidating and discouraging long-armed presence in the key, and an avid collector of rebounds; he was not acclaimed for his shooting accuracy from the field or from the free-throw line.

"I was always a terrible shooter in basketball," he states flatly. "I still have a terrible hitch in my shot today. But it was the countless hours of training in the gym that made the moment possible."

Time was running out. In basketball parlance, the shot clock was about to tick to zero as Josh drew a careful bead on the top of the monument. And then, just before losing his tenuous foothold, as the support structure of classmates crumbled again into a greasy, tangled spaghetti-bowl of arms and legs and heads and torsos and blue Navy shorts, he tossed the cap, carefully, gently, but not *too* gently.

"As I fell away, from the cheers of my classmates I realized the cap had stuck," Josh remembers, "it stuck in the lard, at the top the monument."

And the 1996 Herndon Climb, for the class of '99, was completed. The official time, recorded for posterity, was two hours, eight minutes, and forty-five seconds, with Joshua Caleb Williams earning the singular title of "scaler" for his class. Never was such a title more richly deserved.

It was all over, except, of course, for the celebration. Cheers and shouts of joy echoed across the Academy grounds, along with a few dozen repetitions of "Plebes no more!" and a great deal of accompanying fist pumping and chest thumping. A solemn, heartfelt chorus of the alma mater followed in turn, punctuated with its exclamation point of "beat Army!"—a sentiment that is never far from the mind of a midshipman.

Virtually no one skates through a plebe year. While every class might insist that they had it tougher than the class before or after them, it seems that no one worked harder than the class of '99. Dropouts and expulsions plagued the class at a significantly higher rate than is the norm; rumors flew that a congressional inquiry would be launched if just a couple of more class members punched out before completing the year.

Josh attributes at least part of this to an unusually hot, humid plebe summer, and perhaps to a broken thermometer, which is used to determine not just the ambient temperature but also the amount of physical stress that is deemed acceptable to place on the freshmen as they proceed through those first few torturous months of intensive physical and psychological training.

The "bulb test," Josh and others suspect, was likely malfunctioning, failing to identify certain "black flag" days, where the heat and humidity soar to such a level that the physical demands of training must be eased off to reduce the risk of filling the infirmary with cases of dehydration, heat exhaustion, or worse.

"The color of the flag determines what level of physical stress or punishment can be endured and dealt out by the upperclassmen," Josh explains, "and on a black flag day, there are certain things they can do and certain things they can't do." And there were a number of days, according to Josh and others, that were deemed red or white flag days even though the consensus of opinion argued, with futility, that they should have been black flag days.

Now, following the successful completion of the Herndon Climb, the remaining plebes were relieved and proud to find themselves at the very end of an undoubtedly brutal academic year. Only one more task needed to be accomplished, and it was a relatively simple one for the class of '99: all they had to do was carry the six feet eleven inches and 230 pounds of Josh Williams to the reviewing stand and up the steps of the Chapel, where the superintendent of the Academy, Adm. Charles Larson, waited, ready to present him with a pair of four-star admiral's shoulder boards, representing the expectation that Williams would be the first member of his class to achieve that exalted rank in the Navy.

"Getting hauled out to meet the admiral, on the shoulders of my classmates, was a truly amazing feeling," Josh recalls. Like most of the scalers over the years, he never expected to be the first one to get to the top of the pile in his class' Herndon Climb, and just as is the case with every other scaler, Josh did not achieve flag rank. "At my size, I had it in mind that I would lift the 'First Admiral' to the top . . . and while I always thought it would be pretty awesome to be the first admiral of my class, I realized it was not in the cards, as my size was not ideally suited for shipboard life."

Ironically, due to the expediencies and timing of his service duties, Josh eventually received his honorable discharge from the Navy a few months ahead of schedule, so he was likely one of the first members of the class of '99 to return to civilian life.

Nowadays, Josh is busy running a successful business and raising a family of four children: three daughters and a son, all of whom, not surprisingly, stand slightly above average in height.

And that pair of ceremonial admiral's shoulder boards is mounted and framed, proudly and prominently displayed in his Michigan home, alongside a framed copy of the front page of *The Capital*, an Annapolis daily newspaper, from May 21, 1996. He has his parents to thank for those souvenirs, which they showed proudly in their own home throughout the entire length of Josh's career at the Academy and subsequent service as an officer in the United States Navy.

The May 21, 1996, issue of *The Capital* is fronted with a photograph that captures Josh's first assault on the monument. The photo shows him precariously balanced upon the scaffolding of Derek and Seth and fellow hoopsters and classmates Skip Victor and Ryan Sturgill as he pulled off the Dixie Cup, a split second before the scaffold collapsed into oblivion.

Josh is justifiably proud of that moment and that photograph, which tells a story he will never forget. He is philosophical as he ponders his starring role in the incredible drama of the 1996 Herndon Climb with the class of '99.

"What I get from this is how futile is effort when individuals work alone and unorganized," he observes. "It took us over two hours to complete the task on a sweltering and humid day in May. While we had all our strength and vigor, we were an exercise in futility and chaos. As I look back, I wonder if we could have finished the task in about thirty seconds if we had started with the same singular focus that we ended with."

Josh's unique accomplishment in the Herndon Climb ranks highly among the most special memories of his Academy days. In fact, most of the men and women who attend the Academy point to the Herndon Climb, along with graduation (or "commissioning" in Navy parlance), as singularly proud moments in their lives. For Josh, they may be secondary to his successful basketball career.

In four years under coach Don DeVoe, Josh was a strong contributor on a series of excellent teams. During those seasons, the midshipmen generally had their way against Army and other Patriot League teams. Josh points with pride to the four Navy stars he collected as a member of teams that consistently defeated their West Point rivals. Overall, Josh compiled a dominating 9–1 record in games against Army.

"Some of my best dunks were on national television when we played Army," Josh says with a chuckle. "It was always great to post up against guys who were four or five inches shorter than me."

In his sophomore and junior seasons, the midshipmen fairly dominated their Patriot League opponents, resulting in automatic invitations to the “Big Dance,” otherwise known as the NCAA Championship tournament. In the 1997 tournament, the Navy team drew a number fifteen seed, matching them against a powerful second-seeded squad, the Rick Majerus–coached Utah Utes. Utah was led by Keith Van Horn, certainly one of the toughest and most talented frontcourt performers Josh would ever face. The midshipmen held their own but ultimately lost in that first-round contest by a score of 75-61.

The following year, after again waltzing through the Patriot League tournament, the men of the Naval Academy drew a second consecutive automatic NCAA bid, sending them into the midst of “March Madness” and pitting them against the top-seeded Tar Heels of North Carolina, under head coach Bill Guthridge. That hypertalented Carolina squad boasted future NBA All-Stars Antwan Jamison and Vince Carter, and the midshipmen found themselves overmatched, eventually winding up on the losing end of an 88-52 outcome. Despite the lopsided final score, Josh carries fond memories of that game, especially one particular play where he delivered a surgically precise elbow strike to Jamison’s chin, knocking him to the floor.

“It was accidental, of course.” Of course! “He was so much faster than me, he just came out of nowhere,” Josh explains.

It was all in a day’s work for Josh Williams, one of the true legends of the Herndon Climb and a man who left an indelible mark not only on the face of an NBA legend, but in the continuing lore of the Academy.

10

THE HEIGHT OF GLORY, PART 3

(or, From Agony to Ecstasy)

In war, there is no substitute for victory.

—Douglas MacArthur

A DOZEN OR MORE BODIES tumble down, peeling away from their unstable perches, defeated by the immovable granite. Among them are the classmates who were standing on your shoulders and are now plummeting into the crowd below. It looks like the entire third ring on your side has vanished, along with a good number of second-level personnel, and a great deal of space has opened up between the bodies adjacent to you. This is a major structural collapse, accompanied by the familiar mighty groan of disappointment from the crowd of spectators, this time accompanied by genuine screams of terror. Somewhere in that crowd are your parents, and although you haven't seen them in the sea of faces and you've no idea if they have spotted you, if they saw you fall from a height of nine or ten feet or more, you know that they would have been screaming too.

Now, having matched yourselves against this task and suffered a setback, you are better prepared to renew the struggle.

Hopefully that preparedness will be enough to overcome the pain and fatigue that are beginning to wrack your entire body as the second and third levels of plebes quickly reassume formations, ready to resume the battle.

The faces are different, but the bodies are very much the same. Except for the hulks below you in that solid bottom layer, who stand as firmly as if they are an extension of Herndon's hard granite, the men and women around and above you are mostly interchangeable, all fighting with every ounce of the strength and resolve that enabled them to survive a plebe year, all committed to the same goal: bringing that plebe year to a final, officially sanctioned end. But these bodies are more muddy, more sweaty, more slick with perspiration and greasy flecks of grass than the ones they replace. Unlike you, who climbed onto the second level of bodies while you were still relatively clean and unscathed, these fresh troops are veterans of hard duty in the trenches below.

Two dirty, muddy, greasy feet are now planted on your shoulders, you attempt to reestablish the sense of balance that you felt before, and you strain to support the weight of the newly staffed third layer. The mud is a relatively new addition to your experience today, and it is the inevitable result of the soaking that the grassy lawn around the base of the monument has absorbed from the hoses, wielded by third-class midshipmen. Since the very start of the Climb they have been spraying steady streams of water over the participants, ostensibly in an effort to cool them in the sultry May heat and give them a source of moisture for their dry throats, replacing a tiny bit of the fluid that has been lost in perspiration. But as a side effect, the ground at the base of Herndon has become a soggy mud pit, and the condition is worsening by the moment. Those heavily burdened guys in the bottom layer remain solid and unyielding, but adding to their discomfort is that they are now ankle-deep in mud.

The foot on your left shoulder is sock-clad, and the other, belonging to a different classmate, is bare. Both are caked with mud. For a moment you ponder the strategic value of wearing socks versus removing them. You have removed your own, but now you wonder if the coarse, military-issued socks might offer better traction than bare feet. The downside is for the person who is feeling the results of that traction on sore shoulders: like being massaged with heavy-grit sandpaper.

The packed crowd of muddy, sweaty, straining bodies is beginning to produce a noticeable, not entirely pleasant aroma that at first just tickles, then assaults, your nostrils. It is not an unfamiliar smell; during the hot and humid summer before the academic year commenced you became well-acquainted with the phenomenon known as the plebe funk: a heavy, soupy stink-fog resulting from the exertions of sixteen-hour days of intense physical training by hundreds of young men and women, alleviated only by five-minute showers beneath trickling showerheads of tepid water. The funk seemed to settle over the Yard, and it penetrated every room in Bancroft Hall, where the close quarters only increased the humidity of your environment.

Now, memories of the funk are rapidly diminishing as you realize it was like a fine potpourri of incense compared to this new, awful olfactory sensation. The smell of Herndon is like the plebe funk concentrated a hundredfold, with the added aroma of grease that has been simmering in the hot sunshine for several hours. Whether the grease is lard or Crisco you still don't know, since you became separated from your baking-expert roommate in the initial rush to the monument.

Whenever you dare to turn your head a little bit to the right or left, trying to find a place where the air was freshest, your nose is inches from a dirty, straining foot. As you glance upward, you realize that whoever promulgated the myth that women's feet don't generally smell as bad as a man's never met your classmates or participated in a Herndon Climb.

The minutes have stretched into an hour or more. Keeping track of the precise time is impossible and unhelpful, although you sense a resounding cannon-blast regularly. It fires every ten or fifteen minutes, you can't recall which, and you aren't even aware of the number of blasts you've heard from that cannon, a greatly miniaturized, almost toylike howitzer, the same one they use at the Navy football games, firing it each time the midshipmen score a touchdown. You wish you had heard it more frequently during the football season, a middling one for this year's Navy gridiron team. Now it is nothing more than an unnecessary and disturbing reminder that the Climb is wearing on, and on, and on.

It can't last forever, but it's beginning to feel that way.

You rally yourself and dig a little deeper into your reserves of strength, aware that bodies are still climbing and still tumbling down, although there hasn't been another significant structural failure on the scale of the earlier one. That disaster resulted in a rebuilt structure that feels significantly stronger and more stable, despite the fatigue that shows plainly on the grimacing faces and in the trembling limbs of your classmates.

The cannon booms again. And again the crowd of spectators is hushed for a moment, startled by yet another report from that pesky, noisy little mini-howitzer.

You hear the sounds but you do not use your mental faculties for such pointless efforts as keeping track of time. The ordeal will end when it ends; you must simply hold on, draw inspiration from your equally tasked classmates, and push through to the end.

How much do you want it? That is the question on everyone's minds, and on some of their lips. You hear the shouting. "How much do you want it?" is called out to inspire each other. "Let's go! Let's go! How much do you want it?"

Hose water and sweat and grease trickle down from above you as everyone braces themselves, pushing upward, hoping that somewhere on one of the four sides of Herndon, someone is *close* to the top. The breadth of your view on the overall progress of the Climb is severely restricted, but there are shouted rumors, intelligence passed around and from above you. "They're almost there!" and other optimistic projections are most frequent. You redouble, triple, quadruple the effort to reach ever-depleting depths in your reserve stores of strength and endurance.

The Herndon Monument was dedicated in 1860, and it soon became the focal point of the long-standing tradition that defines the end of a midshipman's first year. *U.S. Naval Institute photo archive*

Second-year midshipmen, or "youngsters," apply a heavy coating of grease to the surface of the monument in preparation for the Climb in this undated photo. Note the long strips of tape near the top, used to secure the Dixie Cup hat and provide a further obstruction to the climbers' success. *U.S. Naval Institute photo archive*

Midshipman Samuel G. Underhill, class of '57, hauls himself to the top of Herndon in this 1954 photo. *U.S. Naval Institute photo archive*

1959: Plebes from the class of '62 successfully scale the monument, placing not one but two combo covers at the top, an emphatic statement of "No more plebes!" One class member displays a sign, only partially legible, that reads "Woe to me if my firstie is [illegible]." *Official U.S. Navy photograph*

Left: Kenneth Grimes, class of '89, tops the monument after a respectably quick Climb that lasted just one hour and twenty-three minutes. *U.S. Naval Institute photo archive*

Below: The Climb begins for the class of '96. The youngsters who prepared the monument left a taunting message in their greasy handiwork: "NO MORE CHITS 96," meaning no more excuses. *U.S. Naval Institute photo archive*

Jubilation is apparent in the faces and body language of spectators and classmates alike as Samuel Underhill of the class of '57 reaches the top of the monument. Note that the standard of dress for the spectators—including coats and ties for men and hats and gloves for women—differs considerably from what is seen in modern times. *Official U.S. Navy photograph*

Vice Adm. William P. Lawrence, superintendent of the Academy, presents Herndon scaler Andreas Bierbrauer, class of '84, with an honorary pair of admiral's shoulder boards. The shoulder boards are ceremonial, but the proud smiles, firm congratulatory handshake, and clumps of residual grease from the just-completed Climb are genuine. *U.S. Naval Institute photo archive*

William Lewis Herndon (1813–57) was a commander in the U.S. Navy, captain of the SS *Central America.* Determined to save as many lives as possible at the cost of his own, Herndon refused to abandon his ship, even as it sank beneath him. *U.S. Naval Institute photo archive*

Members of the class of '63 are eager to shed their uniform jackets as well as their plebe status in this 1960 photograph. When the Herndon Climb took place after graduation for the seniors, unlike in recent years, plebes were dressed accordingly in uniform. *Photo by John Neubauer*

Plebes of the class of '63 are confronted by a sign, left by the youngsters of '62, mocking their resolve and fortitude: "There Is More Grease on This Monument Than There Is in All of 63." *Photo by John Neubauer*

Left: In this 1965 photo, plebes of the class of '68 perform an impressive high-level balancing act to reach the apex of the monument. *Official U.S. Navy photograph*

Below: Plebes of the class of '69 participate in a very muddy Climb. One plebe wears a T-shirt that says "Punahou," perhaps a reference to Punahou School in Hawaii, while a classmate has written the inspirational words "Beat Army" on the back of his shorts. *Official U.S. Navy photograph*

In this 1971 photo, the humble Dixie Cup, symbol of plebe status at the Academy, seems well out of reach for the class of '74. Note the participant at center left, who has proudly written "74" on the back of his shirt in large letters. *Official U.S. Navy photograph by PHCS Ken Bumpus*

The strain of effort may be seen on the faces of these mud-covered plebes of the class of '79. This was the final Herndon Climb before the admission of the first women into the Academy later that year. *U.S. Naval Institute photo archive*

Left: Plebes reach their moment of triumph in this perfectly timed but undated photo. Note the individual at the top clutching the Dixie Cup hat in his hand, while below him a classmate passes up the combo cover to be placed at the top of the monument. *U.S. Naval Institute photo archive*

Below: The class of '86 begins the Climb, confronting a heavily greased monument, which bears the signature of the '85 youngsters. Somewhere in the crowd, against the base of the monument but not visible in this photo, is James McNeal, coauthor of this book. *Photograph by* Baltimore Sun

Kristen Dickmann (1989–2008, class of 2011) demonstrates the sense of humor that made her a beloved member of the class of 2011. Her classmates dedicated their efforts in the Herndon Climb to her very special memory. *Photo courtesy of Carrie Dickmann*

The Herndon Climb is in progress, but still has a long, long way to go until Steve Fortmann's memorable heroics atop the monument. The class of '98's recorded completion time of four hours and five minutes is a record in futility that still stands, but they wear this distinction as a badge of honor. *Roger Miller, photographer / © 2019 Roger Miller*

Greg Reichel, class of 2011, raises his fist triumphantly after placing the precious combo cover that belonged to his late classmate Kristen Dickmann on top of Herndon. *Photo courtesy of Greg Reichel*

Pamela Pitkin is all smiles after the class of '82, only the third Academy class to include women, completes the Herndon Climb in 1979. *Photo courtesy of Pam Pitkin*

Matt LaBonte (*center*) and other plebes of the class of '93 demonstrate the typical finery worn by participants in the modern-era Herndon Climb. *Photo courtesy of Matt LaBonte*

Above: Chris Bianchi, Herndon scaler for the class of 2019, shows his pride in his country and the Navy, along with the elaborate image tattooed on his back, a tribute to his father, Kevin Bianchi, class of '85, who perished in a tragic accident during a routine helicopter training mission. *Photo courtesy of Chris Bianchi*

Right: Lt. Cdr. Kevin Bianchi (1963–2003), with his wife, Barrie, and young son Chris, who would follow in his late father's footsteps and enter the Naval Academy in 2015. *Photo courtesy of Rita Barrie Bianchi*

Barbara Ives was among the first women to enter the Academy with the class of 1980. Although she was unable to place the combo cover, this sequence of photos shows that Barbara was an active participant in the Herndon Climb for her class. *Photo courtesy of Barbara Ives*

Suddenly the noise of the crowd begins to swell, a rising crescendo, rapidly increasing in volume and pitch. *Something* is happening, but your view of the proceedings is far worse than the average spectator's.

All you can do is bear up, dig even deeper, and stay strong with your classmates. You've heard the happy throngs of spectators make noise before. One of those occasions ended in disaster, and all the rest in simple disappointment. But this time it sounds different, louder, even more infused with optimism, and for a second your mind processes that rare moment when you experience the unique feeling of joy and pride that only comes when everyone is rooting for you.

Then, it happens. A massive cheer erupts from the crowd, and within a second, the news of success wraps around the monument. Some of your classmates are falling from the pile that you've built and maintained at the cost of every ounce of strength you possess, but this time they are leaping down purposely, sliding down ecstatically.

A moment later, the joyful chorus begins, sung out by all the massed spectators as well as the participants: "Plebes no more! Plebes no more!" On it goes, for about twenty repetitions, followed by an encore of the proudest version of the alma mater you've ever heard or sung. For good measure the traditional punctuation of the song, a resounding "beat Army!" comes across with a little bit of extra enthusiasm.

All around you, happy chaos and celebratory confusion reign. You want to celebrate but there is nothing in the book of regulations to guide you, so you and a classmate from a different company, whom you don't know and you've never met, engage in a spontaneous embrace.

You see the "scaler," the class member who successfully placed the officer's combo cover at the top of the monument. He is being borne aloft on strong shoulders, to be carried up the steps of Zimmerman Bandstand and personally congratulated by the superintendent and various other gathered dignitaries. You don't know your scaler, again it is someone who is not in your company; but soon you will know his name, and he will gain a semilegendary status among your classmates.

Somehow, you locate your parents. They want to embrace you, but you recoil slightly because you are covered in mud and sweat and grease and you don't want to get them covered in that peculiarly pungent mixture.

But they point out that they've seen you in worse shape, sometime in your childhood, and that a good laundering or dry cleaning should erase all fluid traces of the Climb from their clothes. It's a small price to pay, they reason, and you cannot disagree. So the embrace is made even more special, and entirely unique.

It is one of the proudest moments of all of your lives.

You embrace your parents, and a score of friends and companymates, and their parents, too. You overhear someone ask, "What was that stuff all over it? Crisco, or what?" and you still don't know the answer.

The celebration spreads over the Yard. As the fatigue and pain of the Climb miraculously fade away, cameras seemingly materialize out of thin air. The Tecumseh statue, with its own rich history and significance as a good-luck symbol for generations of midshipmen, is a popular backdrop for photos of newly liberated former plebes.

Flatbed trucks heavily laden with rolls of sod arrive, and they park on Blair Road, spoiling any attempt at a photographic composition of Herndon with the visually striking Chapel in the background. As the area immediately around the monument begins to clear, the real heroes of the day—those of the groundskeeping crew—move in. Within a few hours, the monument will be power-washed to its pristine state and the enormous mud pit at its base will be replaced by fresh, green grass.

Right up the front steps of Bancroft Hall and beyond, there is nothing but joy, happy smiles, embraces, and high-fives. It has been a day of historic firsts in your life, and now you experience another, as you and a classmate are chatting excitedly. A group of uniformed upperclassmen approaches you, and instead of demanding that you recite some obscure nonsense from *Reef Points*, as you've come to expect throughout the year, they offer you their congratulations. And each one in turn grasps your hand in the firm, sincere handshake of a Navy officer.

Finally, you make your way to your room. You have been granted an evening of town liberty for the first time as something other than a plebe: a true midshipman, with rights and privileges you've been dreaming of since your first day at the Academy. But first, your thoughts turn to a shower.

Among your roommates you are the first to arrive, so they will be facing a long, impatient wait for their turn as you wash the muddy, greasy residue of the Climb from your body.

That residue may flow down the drain, but the proud memory of your accomplishment, in completing your plebe year and successfully accomplishing the Herndon Climb, will remain with you for the rest of your life.

11

KRISTEN'S STORY

(or, The Greatest Year of Her Life)

And we wept, that one so lovely should have a life so brief.

—William Cullen Bryant, *The Death of the Flowers,* 1825

Perpetual optimism is a force multiplier.

—Colin Powell, *It Worked For Me: In Life and Leadership*

Think of the brightest, friendliest, most likeable teenager you've ever known.

Smart and personable, always smiling, quick with a kind and funny word of encouragement, existentially positive.

Maybe it's your son or daughter, or one of their closest chums. Maybe it's a nice kid from around the neighborhood, a babysitter, perhaps, or a family friend or relative.

Good at sports. Gets good grades. An overachiever. The very personification of the ideal Boy Scout or Girl Scout.

You've taken for granted the smile, the friendly and respectful word of greeting with every interaction, whether those are frequent or infrequent, and whether or not the relationship is particularly close. You just *know* that's a person who will accomplish something good, maybe something great. That's a person whose very posture and appearance demonstrate an inner sense of honor, of pride, and of compassion.

Now, imagine one day that that young man or woman was taken away, gone without the slightest warning. The picture of health, in the prime of life, simply *gone*, never to return again.

Welcome to the corridors of Bancroft Hall on May 5, 2008, the day that Kristen Dickmann did not wake up in the morning.

If you asked her why everyone called her Dink, according to her closest friends and family, Kristen was most likely to attempt to explain the origin of her nickname by happily bursting into song.

"Skid-a-ma-rink, a dink, a dink, skid-a-ma-rink a do, I love you!" So go the lyrics to the popular tune, beloved by a certain generation of children, and tolerated by parents, since their earliest preschool music sessions.

Even after she took the oath and entered the United States Naval Academy, the song was on her lips as frequently as the alma mater.

"We called her Dickmann at the Academy, because like everyone else you get the last name, unless you have a common last name like I do," says Harrison Smith, one of at least 2.4 million other Smiths in the nation, according to the United States Census of the year 2000. Harrison, or Harry to his friends, knew Kristen very well, serving as squad leader for the second half of her plebe summer and her second academic semester, when he was a senior with the class of 2008. Nevertheless, he was aware that her well-attached nickname among family and friends was Dink.

And he remembers that Kristen impressed him as an exemplary midshipman and representative of her class.

"She was an all-star. During plebe summer she was obviously a leader in her company, for the plebes," Harry recalls. "She had a very sarcastic sense of humor. It was dry, but it was witty. And during the academic year, she shot away . . . a stellar student as well as a varsity athlete . . . a freshman on the [women's] varsity volleyball team, as a plebe. One of the things that she always did was—I know it's a cliché—but she had a couple of classmates who were probably more like me, like the dumb rocks of the company, and she would privately tutor them, just to make sure that they were doing okay, on top of her normal load."

As Harry describes his memories of Kristen, specifically of the day he heard the stunning news of her untimely death, his tone shifts from one of self-deprecating humor to a melancholy sadness that will probably always remain with him.

"I was home on leave, in St. Louis with my parents," Harry explains. "We were getting ready to go to lunch with one of their friends, and I got the call. It was definitely surreal."

Having heard the terrible news that Kristen Dickmann, the bright and shining star of her class and company, had been pronounced dead of a cause that was yet to be determined, Harry could do nothing but catch the first available flight back to Annapolis.

"I wanted to be there for my squad, as well as her roommates," he says, "and then, trying to take care of all the things that had to be taken care of... it was pretty hard for the squad. She was definitely a central member of our group. And packing up her stuff and talking to her parents . . . that was definitely a tough time. Everyone had to be in shock. . . . She was a varsity athlete as a plebe, so this is someone who's in phenomenal shape . . . just an absolute shock."

He firmly believes that Kristen would have graduated from the Academy near the top of her class of 2011 and gone on to a successful career as a Navy officer had tragedy not intervened.

As a plebe, she was not yet obligated to commit to a specialty, but her intended and oft-expressed choice perfectly reflected her positive nature and sunny personality. Kristen dreamed of flying C-2 Greyhounds, the big twin-engine planes that transport cargo between aircraft carriers at sea and shore-based installations. But why would a plebe have her heart set on such a specific goal, to fly a particular type of Navy aircraft, one that seems more obscure and mundane than glamorous and heroic?

The answer is that she had an interest in one, specific type of cargo that is generally tasked only to the C-2 and its successors: personal mail deliveries.

"Her reasoning was that she wanted to deliver mail to the ships, because mail makes people happy. That was Dink," Harry explains. "She was upbeat. I mean, *always* upbeat, no matter what was going on, which is pretty remarkable. If she ever had a bad day, she never let on. I never saw it in the year that I spent with her."

Weeks after her sudden and mysterious death, questions lingered as to how such a tragedy could possibly befall a person in such apparently vigorous good health as midshipman Kristen Marie Dickmann, whose fitness had been seriously tested on an almost daily basis during her year at the Academy.

Those questions were finally answered on June 30, nearly two months after her death, when the Anne Arundel County Coroner's Office released its final report on their investigation. The report concluded that Kristen was the victim of a previously undiagnosed heart arrhythmia, which can often result in the sudden cardiac death of an individual who is otherwise healthy.

This news came as little solace to those who loved Kristen, her family and friends who knew her as Dink. And it came too late to be of any consequence to her classmates, who still had work to accomplish despite the painful loss of one of their best and brightest and most inspirational. They chose to honor Kristen's memory and legacy at Herndon with a simple gesture that represented a completely unprecedented break from tradition.

This year, for the first time in history, the officer's combo cover that would ultimately top the monument at the conclusion of the Climb would be the type of peaked cap designed to be worn exclusively by female officers. More specifically, it was the one that actually belonged to Kristen. And the plebe cover that would be placed atop the monument before the Climb, the common cap affectionately known as a Dixie Cup, would also be one of Kristen's own.

As a sophomore in 1st Company, Dan Hanley and his mates were charged with the task of greasing and otherwise preparing the Herndon obelisk so that it would be ready for the class of 2011. Dan recalls the personal warning he received from his 1st Company officer, Maj. (later Col.) Michael Styskal.

"He handed me [Kristen's] plebe cover and said, you are not to screw this up," Dan laughs. "He made it very clear to me that I was not to mess up that part of it, because of the meaning it had for the class of 2011."

Dan already knew something about sacred trusts. As a plebe he was selected to join an elite and honored unit of midshipmen: Team Bill, which is tasked with the care and feeding of the Academy's happy herd of proud mascots.

"I grew up in Texas, and I raised goats in high school," he explains, "and one of the upperclassmen knew that I raised goats and he came to me one day and said, hey, a farm is donating Bill 33 and Bill 34, will you come over and make it look like they're donating goats to knowledgeable people?"

As for greasing the Herndon Monument itself, Dan and his company mates made no allowances for an easy Climb, sparing no effort to make it just as difficult for the plebes as his predecessors had done for his own class the prior year. They started working late at night and continued into the wee hours until the monument was thoroughly coated with nearly a hundred pounds of lard.

"We went out at about ten, eleven o'clock at night, with a ladder and two fifty-pound boxes of lard from King Hall, and just started coating the whole thing," Dan recalls. "We tried to smooth it out as much as we could, but then we had a lot left, so at that point we were just making grease balls and splatting it on. It was probably three or four hours. I didn't hit the rack until about three in the morning."

As for the Climb itself, Dan didn't witness a single minute of it.

"I was exhausted, and I slept in, and missed the whole thing," he laughs.

Cate Sheerin vividly recalls the terrible, tragic events of May 5, 2008. But she prefers to reflect upon her most fond memories of Kristen, her friend and second-semester roommate.

"She was the absolute best. She was a light, a force of comic relief during plebe summer, just because of her nature," Cate says. "She was just a really funny and outgoing person, not somebody who easily conformed to the plebe mentality and the Academy's rules. She was just so funny. She had those big beefy cheeks that made her blue eyes just pop. Skinny as a rail, so her uniforms hung off of her tall, volleyball-athletic frame."

On the rare occasions that they were allowed to listen to music, it was always country music, of which Kristen was most fond and to which Cate stood in need of an introduction. When country music superstar Brad Paisley and his band performed a concert at the Academy, it was a can't-miss event for both Cate and Kristen.

"I was so excited about the concert, I forgot to wear my name-tag, and I was out of uniform," Cate recalls. "Kristen was making fun of me the entire time we were walking from our room to Alumni Hall for the concert."

Rail-thin and something of a fussy eater, Kristen didn't care much for the food served at daily meals in King Hall, according to Cate. "Her parents would send her packages of trail mix . . . raisins and peanuts and M&Ms, and that's all she would eat sometimes. She'd pick out the M&Ms and eat those straight."

Musical and dietary preferences aside, Kristen's goal-oriented mindset and sense of purposeful direction were what impressed Cate the most.

"I had no idea what I wanted to do in the Navy as a plebe. I was all over the place. She *knew* she wanted to be a pilot, like her father. I always admired that about her, how she was so sure about what she wanted to do when there were so many different opportunities available," she explains.

Tragically, those sky-high ambitions were cut short, and the first person who knew it was Cate, who returned to their room prior to noon meal formation to find that Kristen was unconscious and nonresponsive.

"I was talking to her, and talking to myself, and I noticed that she wasn't responding," Cate recalls. "So I jumped up on her rack, and as soon as I saw her face, I knew. Knowing that I didn't think she would wake up . . . it just really changed my entire life."

Later, Cate was among the throng of mourners, midshipmen and classmates and family, who packed the Naval Academy Chapel for Kristen's funeral ceremony, and she joined the procession to the cemetery at Hospital Point. In that ceremony, she delivered a eulogy for her late friend and roommate. "But I honestly don't remember what I said. Like I just blocked it out," she says sadly. But she remembers what she saw as Kristen's casket was being lowered.

"The entire volleyball team was there, and you could see all of their tears dropping onto the casket. It was . . . really powerful," Cate recalls. "And then 21st Company, every year we held a picnic on her birthday [beside her grave] at Hospital Point, and we would lay flowers and light candles. Something I thought was really cool is that even now in 21st Company, they always do what they call a Leadership Tour of the cemetery, and they pay their respects and say hello to Kristen and talk about her story."

Aaron Rosa was another member of 21st Company in the class of 2011, so he and Kristen went through the entire academic year together, starting on Induction Day. Their respective rooms in Bancroft Hall were next to each other.

"There are people in your company that you worry about when you go through an experience like plebe summer," Aaron says. "Those people stand out, for a bad reason. Well, Kristen was one of those people that stood out for a good reason. Always optimistic, very happy, not annoyingly so. Always had a very fresh, humorous take on life, which is really crucial during plebe summer."

In one of Aaron's fondest and most vivid memories of Kristen, she was looking far from her best.

"I actually have a picture of this. . . . She's just hanging out in our room, and she has a black eye from boxing class," Aaron says. "I remember that being very jarring for me. Thinking that we go to a college where I don't think twice about a woman coming into my room with a black eye. Talking about how chemistry class was rough, and then she went to boxing class and got her butt kicked. She was that kind of person."

Aaron recalls a perfect spring day in Annapolis, warm and sunny, during the intersessional week when classwork was complete and many midshipmen were still taking exams, while others were beginning to feel a complete sense of relief from their academic burdens and have a few precious moments of something approaching relaxation.

"I remember being really excited for the end of plebe year, and we had just finished academics," Aaron says. "I distinctly remember sitting in my room, and Kristen's room was right next door to mine, and so we were sitting in our rooms. Probably doing something plebe-related. I remember seeing and hearing a commotion . . . and seeing my company senior enlisted leader *run* down the hallway. Full sprint. Weird, because he was in his [summer] whites, or maybe khakis."

The quiet peacefulness of a pleasant May morning was shattered, and the alarm spread quickly. Being plebes, Aaron and his roommates were hesitant to leave their own room, wary of an upperclass stunt that would trap them in one of those frequent, irritatingly meaningless "training" sessions that were such a common occurrence in their daily lives. As they watched a growing crowd around the door to Kristen's room, they were extremely concerned, but they could never have considered the horrific possibility that their classmate was in full cardiac arrest, her roommate and others performing CPR in a desperate attempt to revive her.

"I saw the paramedics come in, and they rushed her off. It spread through the grapevine that Kristen's going to the hospital," Aaron recalls. "She's a varsity athlete, in great shape. I remember thinking maybe she fell out of her rack and broke her arm, because we had some instances of that happening. The last thing you think of is that someone died. You just don't think of that happening."

As the rumor mill churned unreliably, Aaron and his roommates tried to return to normal, and several hours passed before they received any official briefing on Kristen's status. Late in the day, the company was summoned to the wardroom, where their senior chief broke the news to them that Kristen Dickmann was dead.

"I could tell he was kind of in shock," Aaron says. "I found out later that he had taken over from Cate doing CPR. The whole rest of that week was a blur."

Aaron attended the memorial service for Kristen, and shortly after her funeral at Hospital Point, he heard about a suggestion for a precedent-breaking, very special tribute to his fallen classmate.

"Word came down the line. Someone in our class sent out an e-mail proposing that we use Kristen's cover for Herndon," he recalls. "It was in the news that there's never been a female midshipman cover used, breaking the tradition of many years, of a male cover always being put at the top [of the monument]. But I think that was more of a big deal in the news than it was for us. For us it was no question. It was going to happen. That was just the proper and correct thing to do."

Dan Gibbons, class of 2008, was assistant squad leader for the first semester of his academic year as a senior, and battalion commander for the second semester.

"Kristen. Whenever I think of her, I smile," he reflects. "She was a quirky girl, smart as a whip. She was just always smiling. She wanted to make sure that people were having a good time. She liked to joke around a lot, a dorky kind of humor. I don't know how else to say it. Before I was a dad, I would have called it dad jokes. Now it's just called funny."

When he heard the news that Kristen had passed away, he didn't *run* to her room, but he walked purposefully, at something like double speed.

"At that point it was like nothing had happened, like she'd never been there," Dan says. "The bed was made, the desk was as she had left it, it was almost like she was just in class. There was nothing there. She was just gone. That's it."

As battalion commander, he attended her funeral, where he was moved to tears. And he remembers bidding farewell to Kristen with the most reverent gesture of military respect to a fallen officer from one still standing.

"I remember watching the battalion officer, Commander [Joseph] DeLeon. He went up to her casket and gave a slow, solemn salute," Dan recalls. "I decided to do the same thing. And I stood at the casket, and I slowly saluted her, [I had] an overwhelming feeling of gratitude for knowing her, as well as just how fragile life is. . . . It was hard for me to process that she was dead now. That I'm going to continue living, and she's not."

For Dan, it was a slow, sad walk back to Bancroft, his eyes filled with tears, not wanting to talk to anyone in the group of other mourners. Sometime thereafter, he heard of the plebes' determination to honor Kristen with a special tribute at Herndon.

"For these young kids to make a decision that they were going to honor her by having her cover go up there, the first female ever . . . it was just incredible," Dan says. "To throw away [years of] a tradition of placing a male cover on top of Herndon, and finally putting a female cover up there? I was never prouder of those kids."

"I remember one specific play. The ball came over the net, another girl on our team kind of shanked the ball. And Kristen, it was her job to go get that ball, and she went flying after it, sprinting. The word 'stop' was not in her vocabulary. And she ran into the scorer's table, and I remember paper and people and chairs just flying all over the place as she dove into the table. She didn't get the ball, so she just got up and ran, jogged back onto the court like nothing had happened. And we were all just in awe . . . like, what just happened? You just knocked down everything! So, yeah, that girl had a lot of hustle."

Jenny Rudewicz was Kristen's teammate. The two were the only plebes on the 2007–8 Navy women's volleyball squad.

"I remember meeting her during plebe summer," Jenny recalls. "Everybody gets together in Alumni Hall and goes to the sport they want to do. And of course, the recruited kids go to the sport that they've been recruited for. So I met her there, in the seats, in Alumni Hall. And I had the big BCGs, and she had BCGs too."

BCGs, Jenny explains, are the military-issued eyewear for those with less-than-perfect vision, and they are designed for functionality rather than fashion. A prime example of the sardonic humor that is typical of a midshipman, the acronym BCG means "birth control glasses," a reference to their seemingly deliberate unattractiveness.

"I remember walking to T-court for formation, and hearing sirens. Lots of sirens," Jenny says, recalling the tragic day near the end of the academic year, long after the volleyball season concluded. "I remember thinking, wow, I hope everything's okay. Did formation, did lunch, went back to my room to get ready for whatever class came right after lunch, when I got a call from Marissa Watson."

Marissa, class of 2009, was a teammate on that year's squad, and she was calling a meeting. It wasn't a meeting to discuss strategy for their next games. And after "chopping" to the other side of the brigade to the room of another teammate, Jenny found her progress halted by a lieutenant, who denied her entrance to the hallway.

"I said, oh, I just need to go to that room, and I pointed to the second door on the right hand side, and he said, oh, okay," Jenny recalls. The lieutenant immediately understood that she was a member of the volleyball team, undoubtedly a friend of Kristen's, and he knew the solemn nature of the meeting that was taking place at that particular time, in that particular room. "So, I chopped into that room, and most of the team was already there . . . and Marissa said Kristen had to go to the hospital, and she didn't make it."

The sadness and outpouring of grief that filled the small, crowded room served as a catalyst that eventually pulled the surviving team members together, drawing them closer, inspiring them to keep Kristen's memory alive in any way possible.

"I think as midshipmen in general, we live like we're invincible, and it's always sad and surprising to hear when a midshipman or [Army] cadet, or anybody from any of the service academy schools passes away," Jenny says. "It definitely made us strong as a team. We had to come together. She was always with us, and we never forgot her. And I made sure that even after I left [the

Academy] I still told stories to the new girls who were coming in who never met her."

As a longtime fixture in the athletic department and around the Annapolis campus, Mike Schwob has held a variety of coaching and teaching positions, and during the 2007–8 academic year he was Kristen's coach on the Navy women's volleyball team. That year the team made a strong showing in the Patriot League and in nonconference games, compiling an overall record of 21-12, with eight wins against six losses in the conference.

Coach Schwob recruited Kristen out of Unionville High School in Kennett Square, in southeast Pennsylvania. There Kristen was a star player, captain of the Unionville team, selected to the all-league and all-state teams for her senior season.

The United States Naval Academy is not a typical college or university by any stretch of the imagination. As one of the nation's great service institutions, the Academy could offer something the other schools interested in Kristen could not. For many potential recruits, the most basic entrance requirements would be disqualifying, and the prospect of a military lifestyle with its attendant Academy-style discipline was daunting. But for Kristen, whose father and grandfather served in the Air Force and Army, those obstacles were a lure.

If Schwob had any doubts about Kristen as a potential Navy recruit, they were soon erased after he had the opportunity to get to know her as much more than just a dominant force on the volleyball court.

"She played for a strong program [at Unionville], a skilled player, kind of a utility player," he explains. He saw her as Navy-quality material immediately. "There are often times when something draws them. . . . She was one of those special, special people."

The coach remembers hearing ambulance sirens on the yard, which are always concerning but only rarely presage the announcement of genuine tragedy. When he finally heard of Kristen's sudden death, he realized that much of the volleyball team would also soon learn the news through their contacts on the fast-moving electronic grapevine. But to his frustration, he had to refrain from immediately calling a team meeting, as military custom required the family to be notified, in person, by the casualty assistance officer before any such meetings happened.

"It seemed like a very long time between when we found out and then being able to talk to the team," Coach Schwob recalls. "From there, the Academy did a nice job arranging things for our players, grief counseling and things like that. It was the beginning of May, at the time when exams and lots of things were going on."

For Schwob, delivering a eulogy at Kristen's funeral, per the family's request, was "an honor . . . and *the* single hardest thing I've ever had to do in coaching. When things like that happen, when you're going through all of this and there's no book on it, you try and figure it out and navigate as you go."

He is proud of two specifics, touching tributes to Kristen's memory that resulted from the efforts of the Navy athletic department and the women's 2008–9 volleyball team, although he takes no credit for either of them.

For their official team photo, as the season began in the early fall of 2008, the women posed in front of a C-2 Greyhound—the exact "mail plane" that Kristen dreamed of flying—with only the slightest bit of necessary string pulling to borrow the aircraft. And, in the cockpit window, they carefully placed Kristen's distinctive number 6, the jersey she wore during her single season with the team.

Next, the department organized and brought to fruition what they hoped would become an annual Kristen Dickmann Invitational women's volleyball tournament, a preseason event hosted by the Naval Academy, with three other varsity teams invited to compete against the host team. Coach Schwob is proud that the Dickmann Invitational has become a successful ongoing tradition under coaches Larry Bock and Paco Labrador, who succeeded him in turn at the helm of the women's volleyball program.

But before that masterful team photo, and before the inaugural edition of the invitational, the plebes of the class of 2011 planned their own noble tribute to Kristen, and they carried out that plan at Herndon.

As the funeral procession crossed the Decatur Avenue bridge across College Creek to Hospital Point and headed up the hill to the cemetery, a family friend pulled aside Doug Dickmann, Kristen's father.

"She said, 'Doug, this is huge! There's never been a woman's hat up there. Ever!'" Doug recalls. The friend told him about the class of 2011 and their plan for an unprecedented break from tradition at the annual Climb of Herndon Monument. "I didn't realize [the significance of the tradition] at the time."

Doug attended the University of Colorado on an Air Force ROTC scholarship and graduated in 1979 with a degree in aerospace engineering. He went on to serve for six years as an Air Force pilot, flying C-141 Starlifter cargo planes out of McChord Air Force Base, followed by thirty-plus years as a commercial airline pilot. Doug's father was a Korean War veteran and an Air Force crew chief who drove around Florida in a car with a license plate that read KMD11—Kristen's initials and her Navy class—mounted in a souvenir "NAVY" frame, despite his own service in the Air Force.

Doug's daughter was a huge fan of professional and collegiate sports and had a streak of tomboy in her, according to her father. "She was not into toy dolls. I don't remember any dolls around the house. She was blue jeans on, flip-flops, pull your hair back, let's go," Doug says. "Not the type of kid that would, at night, set out what she was going to wear the following morning."

With her father and grandfather as role models, Kristen grew up in northern Kentucky, and later Kennett Square in Pennsylvania, with an interest in flying and in military service. And it was Kristen who chose to follow in those family footsteps, not her older brother, Kyle, or sister, Cassie.

Kyle remembers his sister as an outstanding athlete, interested in volleyball, gymnastics, and dance, as well as a superior student. "She had a wide range of interests and athletic ability, to be able to pick up whatever it was that she was doing and to excel at it, and obviously, be very smart in the classroom also," he says. "It was kind of a given that she would get into the Naval Academy."

As Kyle pursued an economics degree at Vassar College, his youngest sister was an eager helper around the washer and dryer and folding table whenever he came home for weekend visits bearing a college student's inevitable load of laundry. And when his days of building backyard obstacle courses for his delighted young siblings were behind him, he set his sights on a career as a lawyer, earning a law degree from Widener University and eventually setting up a successful tax law firm in Denver, Colorado. If he ever needs to be reminded of his dear sister, he only needs to look at his rib cage, where Kristen's initial is tattooed.

Kristen's closest friend and playmate was her older sister, Cassie, and since the two girls were only two years apart in age, their natural competitiveness drove Kristen to become an advanced learner and a high achiever from a young age. When the two girls weren't heard singing "Skid-a-ma-rink, a dink, a dink" to each other, they might be in the family basement, engaged in play school and challenging each other's skills. As they grew older, both Dickmann sisters were drawn to sports, volleyball in particular, and they drew closer as the new kids in class at Unionville High School. Cassie was a junior and Kristen a freshman when the family moved to Kennett Square.

"We had both started playing volleyball a little later in life, eighth grade for me, sixth for her. We were known as the Kentucky girls, coming to Pennsylvania, playing volleyball together," Cassie explains. Kristen earned a spot on the varsity squad as a freshman, so the two siblings had the opportunity to play together for Cassie's two years at Unionville. With Kristen playing in the back row and Cassie at the net, they occasionally ended up switching spots in the heat of a match and became known as "the Dickmann Switch."

After graduating from Unionville, Cassie went on to play at Juniata College, and when Kristen was a Naval Academy plebe, the two sisters, longtime friends and rivals, were matched against each other. Navy won the game, "because they *always* win," Cassie laughs.

"When she got to the Academy, she chose the number 6 because her sister wore number 3, and she said she was twice as good," says Kristen's mother. Carrie Dickmann was constantly impressed by her daughter's smarts, her determination to never be left behind, and her refusal to be intimidated by challenges, or rules, or authority figures. According to Carrie, her youngest child never allowed herself to be constrained by such limitations as certain school library books that were assumed to be above her reading level.

On her first day of second grade, Kristen had a new teacher who was unfamiliar with the girl's precociousness and intellectual prowess.

"They were all told to go get a book, and the teacher had color-coded them. She told all the second graders to get books off the yellow shelf, while she worked with each one individually," Carrie recalls, "and Kristen went to look for a book, and she came back and interrupted the teacher and said there wasn't anything she could read."

Repeatedly, the teacher sent Kristen to keep looking, even to the next color shelf. Kristen kept looking and kept coming back empty-handed.

"I got all of this when her teacher called me that night, hysterically laughing," says Carrie. "So the teacher finally asked her what kind of books she read, and she said [brusquely], 'I read *chapter* books!' So, the teacher said, go find any book and bring it back and read it to me. Let me hear how well you read. So Kristen went and selected a book and brought it back to the teacher and read it. The teacher said, I am so sorry, I didn't realize . . . and Kristen said to her, 'I didn't think you'd read my file.'"

By the time she got to high school, Kristen's natural leadership abilities, which would suit her well at the Naval Academy, were on full display. Carrie worked at the school, and she recalls the time when she noticed a girl walking out of a classroom carrying a trash can, followed by Kristen carrying two trash cans. Spied by her mother, Kristen merely smiled and kept walking, with one trash can in each hand. Finally, when a third girl emerged from the classroom with a fourth trash can, Carrie stopped her and asked what in the world she was doing. Exasperated and panicking, the girl blurted, "I'm going with Kristen!" Upon investigation, Carrie discovered that somewhere, somehow, Kristen had scavenged the top of a ping-pong table and had convinced her classmates that they could build a table so they could go play ping-pong.

"And this was calculus class!" Carrie laughs.

"She was very, very talented in math and science," Doug recalls. "After high school she wanted to continue playing volleyball, like her sister, and she wanted to study engineering and mathematics. I remember she could do those Sudoku puzzles. I thought, 'how do you see this?' She just did."

Kristen's desire to compete in volleyball and also earn an engineering degree narrowed her potential options at the college level. The quandary was that she might not have made the grade in a Division I, major-college volleyball program, but most of the smaller schools that might offer her an athletic scholarship were liberal arts schools, lacking strong programs in math and science. According to Doug, he pointed out to his daughter that the Naval Academy offered her the opportunity to do both. But making a commitment to the Navy was a major lifestyle change, and Kristen wrestled with the decision.

"I think she felt that sense of Grandpa did it, Dad did it, somebody's got to do it. She felt a little bit of that," Doug observes, recalling his daughter's mixed emotions.

After she attended a summer camp sponsored by the Academy's athletic department and met Coach Schwob, Kristen's reservations about pursuing a military career were finally assuaged.

Kristen's year at the Academy seemed, to Doug, to come to her just as easily and naturally as life at a carefree liberal arts college. Annapolis was just a two-hour drive from the family's Pennsylvania home, so Doug and Carrie were able to attend most of Kristen's volleyball games and visit her in a few other rare and precious instances.

"We'd get down there as much as we could, but every time, when it was time to go home, there were no tears. We'd look around and see other plebes there, and the tears and crying and all that stuff, but always a smile on her face," Doug recalls. "[It seemed like] this experience, to her, was not the same as for the other eleven hundred kids there. There were always tears, but not with her. It was always just a big hug, see you later, and off she went. It was almost like dropping her off at a regular college."

Carrie recalls the ingenious design of the care packages she sent Kristen, which contained such desirable commodities as gummy candies and her beloved Swedish Fish. Knowing that these treats were likely to be pilfered by upperclassmen who were charged with the duty of mail inspection, Kristen devised a clever sub rosa method to defy them.

"Kristen sent a letter home and asked me to buy boxes of tampons and empty them, except for the top row," Carrie says, "and to pack the candies behind them, and then reglue the tops. Because she said, no one was going to open any of her boxes of tampons. Well, every week Kristen got a fresh box of tampons. I'm sure they thought something was wrong with her," Carrie laughs, noting that her daughter always got the candy, and her clever scheme was never discovered.

The ensuing years following Kristen's untimely death have given the family a philosophical perspective. "I'm very blessed, in that when you pass away and your kids are left behind, you don't see the impact they've had," Doug explains. "But when they pass away first, and people say all this stuff about what an impact they had, you think, that's pretty incredible that *my* child did that. You almost pinch yourself and say wow, did I raise that kid to be that way? Or was that just who she was? Maybe it's a combination. I don't know what the breakdown is. But it's nice to hear people talking about your children that way."

Somehow, word spread that Kristen had been a dedicated follower of local sports in northern Kentucky and Cincinnati, the nearest city with professional teams.

"She was a huge Reds fan and Bengals fan. And a huge University of Kentucky basketball fan," Doug says. "When she passed away, we got a letter from [Billy Gillespie], the Kentucky basketball coach. The quarterback for the Bengals, Carson Palmer, sent an autographed jersey. I don't even know how these people knew that she was such a fan of these teams. Maybe it was in a press release or newspaper article, I don't know."

Doug offers a word of advice—a warning perhaps—to Kristen's prospective biographers. "Just make sure you put two 'N's on her last name," he says, "or her grandfather will come down from heaven and whoop all of us for misspelling her name."

On the day of the Climb, Greg Reichel's head was perfectly clean-shaven.

A few days prior, while preparing for the Sea Trials along with their company and the other plebes of the class of 2011, Greg and his roommate had given themselves grossly exaggerated "high and tight" hairstyles.

"After Sea Trials, the next morning, we found out about Kristen's passing. We started hearing about the funeral arrangements and stuff like that, so I was sitting there wearing a ridiculous high and tight haircut, realizing I'm going to a somber funeral. I felt terrible," Greg explains. "I recut my hair for Kristen's memorial service. So I had a completely shaved head, like completely bald shaven, in all my pictures of Herndon."

Some classes have successfully performed the Climb efficiently and quickly, with sheer luck often playing as important a role as effective organization and communication. Other classes have slipped and slid and stumbled and bumbled their way through two-hour-plus ordeals before the task was finally completed. The class of 2011, standing firmly among the latter, needed two hours and thirty-six minutes to cap the monument.

"Our class had a very long Climb compared to all the historic ones," Greg admits. "We got there, and people started climbing . . . and it just kept collapsing because everyone was trying to get to the top. I actually had zero desire to be on the top of Herndon."

A fixture on the Navy Men's Club volleyball team, Greg stands six feet seven, with long arms that made him not only a terror at the nets but a strong candidate for the honor of being the Herndon "scaler" for his class.

"In the [hours before] Herndon, the news crews were there, and one of them interviewed my dad and asked him what he thought about Herndon, if he knew about the traditions and so forth," Greg says. "And my Dad, being . . . an overly proud parent and six foot seven himself, told the news reporters, I'm here because my son is going to be the one to get to the top of Herndon, because he's six-foot-seven and about 170 pounds, and he's perfect for the job."

What might have seemed like fatherly braggadocio began to look like prescience two and a half hours after the Climb commenced. Greg, however, did not start with the mindset of confirming the wisdom of his father's bold prediction.

"I was just trying to build the base," Greg explains. "I figured my height would help whether I was on the top or on the bottom. I watched the tower fall down, we would climb up and start building the third row, and it would always fall apart because everyone was trying to climb. I spent the first two hours getting my face and my shoulder blades and my collar bones stepped on."

Eventually he realized that something wasn't working, that this Climb was turning into an exercise in near-comic futility. Somehow he worked his way out of the pile, following a major collapse, and surveyed the scene from a new perspective.

"About two hours in . . . everyone was getting just completely gassed and everyone was tired, and people weren't even trying to climb up to those upper levels anymore," Greg recalls. "Someone said to me, you need to go up. You need to be the one to reach the top, because I think we can hold you up. Because I was pretty light. So I started climbing and, because there were fewer people climbing, we were finally able to support one person on the third layer, and it just worked out that I was the one on top. I never once tried to climb above the base level until that moment, because I knew there were just too many people going for it."

So, he managed to get near the top, climbing up the backs and shoulders of his classmates like they were a set of playground monkey bars. There, Greg found the Dixie Cup hat—one that had belonged to Kristen—taped to Herndon's apex with a few strips of duct tape, which he was able to peel away without too much effort. He was well aware of his classmates' determination to honor Kristen by placing her own combo cover on top, and he fully supported the plan, but he had no real idea how it would happen.

"I didn't know who was holding her cover, or where it was, or if someone was holding it in the crowd. Or how that cover was ever going to get into the mix," says Greg. "It just seemed like fate, when I got up there and turned around, and I saw her cover being passed up to me."

As he stood on the slippery shoulders below him, Greg received the cover; but just before he was able to place it on top, once again the supporting human structure collapsed.

"I actually did a free fall, all the way from the top into the crowd at the base," he explains. And then, with Kristen's female midshipman's cover perched on his bald cranium, a determined Greg made his way to the top of the pile a second time. "I climbed back up, on the same people I climbed before."

Finally, triumphantly, Greg placed that very special cover at the top of the Herndon Monument, and the joyful cry of "Plebes no more!" echoed across the yard. It was the first time in history that a female midshipman cover capped the monument, and it would also be the last time, since the Navy officially retired it in 2018, and since then all male and female midshipmen have worn the same "unisex" cover.

It would take some time for the bruises and sores on Greg's shoulders to heal, but the memory that he had played such a key role in the ultimate tribute to a fallen classmate will last a lifetime.

Along with the rest of her class of 2011, Kristen was a plebe no more.

Kristen Marie Dickmann was laid to rest in the United States Naval Academy Cemetery at Hospital Point. There her grave is perfectly well-kept, among the final resting places of such noteworthy heroes and famous names as Sen. John McCain, Adm. James Stockdale, Adm. Stansfield Turner, and Col. John W. Ripley.

Kristen's polished granite headstone is easy to find for well-wishers and visitors who seek to pay their respects to the midshipman whose life was so tragically cut short. Her grave is just steps from the main walking path leading up the hill to the hospital and the Naval Institute, and it is usually adorned with tokens of remembrance; on a typical day ten years after her death, flowers, colored ribbons, American flags, smooth stones, small figurines and carvings, a pilot's-wing pin, a tiny watering can, and a box of Swedish Fish—Kristen's favorite candy—were among the inventory of mementos, some of them obviously bearing a very personal meaning to Kristen and the person who placed them there.

Her headstone is inscribed as follows:

KRISTEN MARIE
DICKMANN

MIDN

APRIL 16, 1989
MAY 5, 2008

CLASS OF 2011
USNA

"PLEBE NO MORE"

Stark and simple, the inscription is a fitting tribute to a young woman who had precious little time on Earth to leave behind a lengthy testament to her courage and honor. But she lived her life in such a way that others who knew her, and served with her, and fought and struggled beside her will always be inspired by her memory.

She left little doubt that her last year, the one year she held the rank of midshipman, was the greatest year of her life.

12

IT HAPPENED AT HERNDON

(or, If Granite Obelisks Could Talk)

Experience is one thing you can't get for nothing.

—Oscar Wilde

The campus of the United States Naval Academy in Annapolis is arguably among the most historically significant educational institutions in the nation. Since its founding in the mid-nineteenth century, the Academy and its denizens have developed myriad customs and traditions and a rich vein of lore and legends. The annual climbing of the Herndon Monument by first-year plebes is one of the most noteworthy of all these customs, with its own historical process of development over the decades, from the relatively low-key "swooping out" of the early twentieth century to the carefully planned and scheduled popular spectacle of the Climb in recent decades.

It has developed its own set of internal procedures and agreed-upon practices, some of which date to the earliest days of the Climb and the swooping out, while others are more recent modifications.

One profound superstition, or myth, that has been widely circulated in the *Reef Points* handbook and by oral tradition is the belief that the midshipman who caps the monument is destined for a brilliant career in the Navy, ultimately to achieve flag rank. And that person is supposed to be the first member of his or her class to do it.

Perhaps no belief has ever been so factually, logically, mathematically, and thoroughly disproven. Since the names of the individual "scalers" were first recorded in the 1940s, none of them has ever been promoted to the rank of admiral. Most, according to the numbers, have long since retired from active duty, which reduces the chances of the myth being proven true just once.

Myths can be stubborn opponents of facts, but the longer they persist, the more strongly they sometimes take hold. The anonymous authors of the 1945–46 edition of *Reef Points* were so eager to promulgate the belief that the Herndon scaler would be the first admiral from his class that they dedicated a few lines of text to an ill-fated and long-scrapped plan that may never have been more than an idle rumor: that a certain patch of ground adjacent to Stribling Walk would be dedicated as "Plebe Park," and a statue would be erected there to the first scaler to become an admiral.

Perhaps the day will come when one of the scalers from recent classes—maybe Christian Schwien, class of 2022, Peter Rossi, '21, Joe McGraw, '20, or Chris Bianchi, '19—will reach flag rank. But the nearest the myth has come to actual fruition was in 1966, when the class of '69 took its turn at the Climb.

Stan Bryant and Jerry Witowski were roommates for all four of their years at the Academy, through graduation, and they maintained an enduring friendship as their career trajectories took them in different directions and across continents as Navy officers and into civilian life. If the two men were somehow fused into a single body, and if their respective, individually impressive résumés were merged into one, that document would show that the Bryant/Witowski entity scaled the Herndon Monument at the conclusion of plebe year and was also the first admiral from the class of '69.

"Is this as close as it has come? One roommate tops Herndon and the other puts on the flag first?" Admiral Bryant asks. "It's a stretch, but can anyone, or two, get closer? I don't know. But it's fun to think that Jerry and I have come as close as it gets to making the legend a reality."

Together, the two friends and lifelong "shipmates" attended the Climb for the class of 2019, fifty years after Witowski scaled the monument and placed the precious combo cover at its apex.

"A lot of things are different now, and many things are very much the same," Jerry observes as he ponders the development of the Climb over the last half-century. "I think it deals with how the individuals choose to attack the problem in the beginning, and how they finally have to attack it in the end to be successful. A common thread is that you learn a lot while you're doing it. I sense that the classes of today are a little better prepared and perhaps a little more organized, with a plan going in. Which always helps but doesn't guarantee success."

In the past fifty-plus years, rapid changes in communication technology and information availability have had a major impact on the culture surrounding the Climb and at the Academy in general. "Before the days of Wikipedia, before the days of the Internet, what we knew about [the Climb] was what we were told by our upperclassmen, which was: this is the plan, this is what you have to do to prepare to climb the monument," says Jerry, who served twenty-two years in the Navy, inclusive of active and reserve duty, finally retiring with the rank of captain.

Media coverage and a general increase in awareness of the Climb have also changed notably, according to Jerry. The event barely merited newspaper coverage in his time, but recent Climbs are extensively covered by local newspapers and television stations and national wire services. It has even been mentioned—albeit briefly—on ESPN's *SportsCenter* as one of the nationally televised cable broadcast's ubiquitous Top Ten Plays of the Week.

"It was a little bit different then, there wasn't quite as much attention," Jerry explains. "I remember somebody tracked me down after the Climb and asked me for an interview. I believe it was someone from either the *Milwaukee Sentinel* or the *Milwaukee*

Journal or an Associated Press stringer, because it was picked up by the newspapers. At the time my parents were living in Hayward, Wisconsin, and they clipped out the one-and-a-half-column-inch story about it."

One aspect of the Climb that has not changed in fifty-plus years and is unlikely to *ever* change is the confidence of midshipmen, toughened by a difficult plebe year, and their determination to face any situation or challenge, knowing that if other classes were successful at Herndon, they will ultimately be successful too. The only factor that changes is how much time will be required to achieve that success. No class has ever failed at the Climb, and no class has ever given up the challenge.

"As I recall, we didn't have a lot of information in advance. It was something that was expected," Jerry says. "No one didn't think we would make it, because every class before certainly had been able to climb the monument. It was just passed on as one of those traditions. And I have a belief that, either at the Naval Academy or in the service, if you do something more than three times, it's a long-standing tradition."

And the mechanics, or mathematics, of the Climb are still intact. The Herndon Monument still stands twenty-one feet high, granite is just as hard and unyielding as it has always been, and grease is still slippery, even though the actual compound selected for the job may vary in composition and quantity. As for the incessant demands placed upon plebes by upperclassmen, that's less a tradition and more a simple fact of life.

"At the end of plebe year, things are still not totally relaxed for the plebes," Jerry explains. "The upperclassmen's attention is somewhat distracted because they've got girlfriends coming, and they're getting ready for June Week, and there are a lot of other activities, so they don't have nearly as much time to dwell on the plebes as they did during the entire preceding year. I remember them wetting the ground with hoses, and they greased [the monument] with something that was just really foul smelling. I have no idea what they put in there, but it made the activity of climbing the monument another unpleasant task, among many that we'd already endured during plebe year."

Jerry's vivid description of his participation in the Climb—in 1966, when the war in Vietnam was raging, Lyndon Johnson was president, and the Beatles were still together, to name just a few of the cultural touchstones of an earlier era—will sound familiar to anyone who has witnessed or participated in a more recent edition of the ritual.

"We went in there knowing that it wasn't going to be an easy climb. My recollection is that [then, as now] the football players and the big guys and the jocks kind of just attack it. Not necessarily with a plan, but they attack it. And the rest of us just swarm around, trying to figure out how we can help. Eventually, nothing works as well as trying to link arms, get the big guys on the bottom, the next guys who don't think they're going to crush the big guys at the next layer, and then a few people try gathering footholds and armholds, but slipping down, and going back in again. This went on for quite a while."

As many successful Herndon scalers have done, Jerry waited patiently for his classmates to form a strong, steady structure before throwing himself into the mix. He points out that his intention wasn't necessarily to get himself to the top, but rather to look for an opportunity to contribute to the overall effort in the most effective way possible. A three-year varsity letterman in football and baseball in high school, Jerry realized that his athleticism was better employed as a climber than as a ground-level pillar of support.

"I'm probably five-seven, five-eight, and at the time I probably weighed 150 pounds, so I was relatively small. Let's just say I was medium sized," Jerry laughs. "I didn't immediately try to get to the top, but I was watching from the perimeter. I tried a couple of times, then went back and watched again, looking for an area that looked like it was accessible to climb, and I do remember a lot of people trying to help. I mean, that's what it was all about."

The Climb is generally viewed as an exercise in cooperation and team building as much as it is a grueling physical challenge for those who throw themselves into the crowd, putting their bodies on the line, and pushing the limits of their strength and endurance. For Jerry, it was all those things, but what impressed him the most about his classmates' effort was the altruism of each individual as everyone worked together to achieve a common goal, knowing that while only one person would get to the top of the monument, they would all share in the reward. The entire class would celebrate together, no one would be left behind, and they would all be "Plebes no more."

As he had anticipated, it wasn't an easy time, but everyone was working together to get through it.

"If they saw you were going to even attempt to get to the second row, you had people pushing you up, and grabbing you up and helping you up," he recalls. "It was a total effort of selflessness. It wasn't a competition, that's for sure. I couldn't even think of anyone else that I was competing with. It wasn't a competition. And there may have been people on the other side [of the monument] that were trying to climb as well, but if someone thought they could help, they would help, in any way they could. Sometimes it didn't help, but they were trying. Pushing and pulling."

The Herndon team-building exercise for the class of '69 finally ended after an official recorded time of an hour and thirty minutes, although it seemed much longer to Witowski, and especially to his roommate Stan Bryant. He was limited to a supporting role, watching the Climb from the perimeter after recovering from a surgical procedure, obeying a doctor's orders instead of brazenly charging into the crush of bodies at Herndon. "You get a little disappointed when you all crumble down the greasy thing, and start going again," Stan recalls, "but the feeling you have when it's over is like no other."

"I put the cap on [the monument] and the pyramid collapsed, almost instantaneously," Jerry laughs. "I had a few scary thoughts about being trapped, as I slid down and all these big guys were in there as well."

In 1966, there was no post-Herndon ritual of carrying the scaler up to the adjacent reviewing stand, either on the Zimmerman Bandstand or the steps of the Chapel, to be greeted by the superintendent and receive a ceremonial pair of admiral's shoulder boards. But Jerry doesn't regret missing out on that particular moment of individual glory. He was quite content to share in the joy of his classmates, and he was mostly looking forward to a long, hot shower.

"We had to move on to the next activity, so everyone needed to get cleaned up pretty quickly for whatever that was," Jerry says. "Someone tracked me down and asked me a few questions, and I don't remember because I was more concerned about getting to the next formation. I saw Stan back in our room and he was as happy as I was. We were all happy because we accomplished the intent, which marked the end of plebe year. So we were no longer required to do things like sit at the meal table a certain way or turn square corners. We were still beholden to the upperclassmen, but in a much more relaxed atmosphere."

His post-Navy career in the private sector gives Jerry a well-informed perspective on the Climb as more than just a simple exercise in teamwork. "Having practiced management and organizational development quite a bit, the whole concept of how the monument gets climbed, I think, is a case study in organizational theory and a few other things about just how much organization you need, how good is the plan going in, what happens when the plan doesn't go off as you hoped, how do you finally achieve the goal, and there's quite a bit of replanning on the fly that goes down in those events," he observes. "It's kind of a microcosm of how certain things get done sometimes. So there are a lot of other aspects to it besides just getting the cap on. It's how you go about it. And what ultimately happens to be successful. There are certain ways you can try all day, and you won't be successful, so you finally have to come around to the right formula."

Regardless of which of them scaled Herndon and who eventually pinned on an admiral's star, both roommates went on to exemplary careers after they graduated in 1969.

Stan Bryant, en route to earning a heavy collection of hardware including a Distinguished Service Medal, three Distinguished Flying Crosses, Air Medal, four Legions of Merit, and numerous other service, campaign, and personal awards and medals. He flew 137 combat missions over Vietnam from the deck of the carrier USS *Enterprise* and commanded the VA-55 Warhorses attack squadron, eventually accruing an astounding 1,058 carrier landings in his career, among other notable accomplishments. He commanded the amphibious transport dock USS *Ponce* and the aircraft carrier USS *Theodore Roosevelt*, leading the *Theodore Roosevelt* to a Battenburg Cup award for operational excellence, the Golden Anchor for retention excellence, and the Battle Effectiveness Award, commonly known as the Battle "E" ribbon. After adding to the visor of his own combo cover a full serving of "scrambled eggs," the golden embellishments that signify the rank of admiral, he assumed command of the Iceland Defense Force operating out of Keflavik Naval Air Station, and later he commanded the Norfolk-based Carrier Strike Group Four. Eventually, Admiral Bryant went on to serve as deputy commander in chief of U.S. naval forces in Europe, finally retiring from duty in 2001, thus concluding thirty-seven years of military service. Nowadays he runs an independent aerospace/military consulting firm and leads training seminars across the country.

Jerry Witowski had an accomplished and distinguished Navy career of his own. He earned his pilot's wings just two years after graduating from the Academy and flew A-6 Intruder twinjet attack aircraft for much of his time on duty, mainly in the Sixth Fleet. After his twenty-two years in the Navy, he went on to a successful tenure in the aerospace industry. "I guess I could say I got to admiral in the company I was working with," he says with a laugh, referring to his position as executive vice president, reporting to the chairman and CEO of an $8 billion company.

Will one of the midshipmen who scaled the monument in recent years ultimately achieve flag rank, or will the next scaler, or the one after that? Will the legend ever be proven true? Only time will tell. But even when it happens, two roommates from the class of '69 will always maintain a special place in Herndon lore.

Somewhere in the cluttered garage of a Maryland home, nestled among a number of nondescript cardboard boxes, is one particular box that contains a rare piece of Naval Academy history. It is one of many, similar-looking boxes, but it should hardly be considered lost or misplaced.

Kevin Delamer, class of '86, knows exactly where to find that particular box with a minimum of searching and rummaging. At any time, he can easily retrieve the precious item stored within, a souvenir from his days at the Academy, which has frequently been displayed on the wall of his various workplaces as a teacher at the Naval War College and the Patuxent River Naval Air Station. Those teaching assignments followed on the heels of a long career as a helicopter pilot, including a noteworthy stint as the lead test pilot for the Navy's VH-92A program, also known as the Presidential Helicopter, designed to provide helicopters for safe, reliable, and timely transport for the president of the United States.

Kevin is understandably proud of that particular souvenir of his Academy days, because it isn't something that anyone can buy at the midshipman's store or the visitor's center gift shop. It wasn't bought, it was *earned*.

"The plaque with the three stars on it." That's how he refers to the mounted pair of admiral's shoulder boards, presented to him in May of 1983 by Vice Adm. Edward Waller, superintendent of the Academy, in the riotous aftermath of the Herndon Climb.

"It wasn't that I had any intention or plan to be the guy who did the hat trick. It just sort of worked out that way," Kevin recalls, "19th Company, we were right there in the front, so we got to the monument just about as fast as anybody. We were all just linking hands and trying to get up. I must have fallen off four or five times.

I ended up with Matt Ware just below me, and we had gotten up to that level where we were just short of being able to do the hat trick. And I literally went from standing on Matt's shoulders to having one foot on his head."

He needed just one more little push, and he got it from Matt, who somehow freed his right hand, and pressed up against the sole of his classmate's foot.

"That allowed me to reach up and yank the Dixie Cup off. They had tried to tape it down with duct tape, which didn't stick very well, with all the grease around. That was fortunate for us," Kevin explains. "Someone passed the combo cover up to me, and then all I remember is being in free fall. Once that cover got up there, the celebration started."

With the start of the celebration came the accompanying collapse of the rickety structure of bodies that Kevin had been standing on. He plummeted from Herndon's height, but he was not injured. His trusted classmates, including several young men who would eventually attain flag rank, not only cushioned his fall but carried him bodily, triumphantly, up the steps of the Chapel to the superintendent's reviewing stand. There, before his proud classmates and the collected multitudes of observers, Admiral Waller bestowed on him the honorary "three stars" of a vice admiral in the United States Navy.

Despite his achievement, and the pride of representing his class as Herndon scaler, Kevin still found himself at the mercy of unforgiving upperclassmen and subject to a good deal of razzing, much of which was not entirely of the good-natured variety.

"I was not the world's most squared-away plebe," Kevin admits, "which made the idea that I was going to be the first admiral in the class somewhat amusing for some of the upper class. I had a couple of seniors who gave me a hard time, who continued to give me a hard time even though I got the cover on Herndon. I remember one of the firsties who'd given me a hard time, telling me, well, you've had your fifteen minutes of fame. I was a little confused, because I was completely oblivious to the fact that my picture had been on the front page of the *New York Times* and the *Washington Post* and the *Baltimore Sun* and the *Capital* (Annapolis), showing the joyous chaos as I was being carried to the steps of the Chapel to get the plaque from Admiral Waller."

But it has been said that time heals all wounds, just as service often improves camaraderie among Navy shipmates. A chance encounter in a Bahrain pub with one of his former Academy antagonists finally brought things full circle in Kevin's relationship with the class of 1983.

"In 1990, I was on the USS *Midway* and we were in the Persian Gulf, during Desert Shield. And it had been decreed that because the Marines in Saudi Arabia couldn't drink alcohol, nobody in the entire theater could drink," Kevin explains. "A squad leader who tried to run me out of the Academy when I was a plebe had gone Marine Air, and he was assigned as a forward air controller to one of the Marine units in Saudi Arabia. I was in Bahrain with my copilot to take care of some squadron business, and we had a two-day layover there. We're at the pub, and there he is, sitting at the bar. And we all know full well the Marines aren't supposed to be in bars or drinking or anything else while all this is going on. I walked up to him with a big [expletive] grin on my face and said hello. I didn't have to say anything else. We both knew it, and I wasn't going to do anything about it, but he knew, hey, this guy can end my career right now. It was just one of those 'who's screwed up now?' moments."

Sometimes, as every child is warned, everything is fun and games until someone gets hurt.

And on the 19th of November 1999, fun and games turned to tragedy. Not only were people hurt—twenty-seven required hospitalization—but twelve people were killed.

Texas A&M University in College Station, Texas, once had a campus tradition that rivaled the Herndon Climb in its long history, the devout pride and sweaty effort of its participants, and the popularity of it as a must-see spectacle.

The great Aggie Bonfire burned once every year, before the important rivalry football game against the University of Texas. It was much more than an enlarged campfire or effigy burning. In the course of its ninety-year history, it began as little more than a burning pile of trash, but over the years it grew by orders of magnitude in size and complexity. In its final years, the bonfire had grown into a cone-shaped wood stack over fifty feet high, a carefully engineered construct of telephone-pole-sized logs wired together and secured with steel cable, requiring months to build, and finally doused with gasoline or jet fuel to ignite the conflagration, which would burn for an entire day and night. When the inferno was burning most furiously, it was visible against the night sky from a distance as far as twenty-five miles.

To the casual observer, most likely not a Texan, the idea of building a structure more than fifty feet high, entirely from wooden logs and wire, dousing it with jet fuel, and setting it ablaze sounds like an accident waiting to happen. That accident happened in 1999, during the construction phase of the bonfire.

In the early morning hours of November 19, as volunteer crews worked to finish building the enormous wood stack, the structure suddenly collapsed. As tons of logs tumbled to the ground, dozens of people were caught in the collapse, and many were trapped and crushed. The rescue effort took more than twenty-four hours and included scores of A&M students and residents of College Station who rallied to the assistance of first responders; members of the Aggie football team were praised for their tireless efforts, laboring mightily, working together to move the heavy logs so that survivors might be reached. Sadly, twelve of the victims did not survive.

In the wake of the tragedy, university administrators decided they could no longer sanction the Aggie Bonfire due to the inherent safety risks associated with the event, primarily the building of such a large structure by mostly inexperienced hands. Organizers and planners of the bonfire had well-established procedures and safety protocols in place—construction of the towering wood stack was by no means a haphazard undertaking—but in spite of their best efforts, disaster struck. Administrators put an end to the bonfire, although the tradition has been maintained in a smaller version, off campus and unsanctioned by the university.

The memory of the tragedy at Texas A&M was still fresh when reform-minded Vice Adm. Jeffrey Fowler assumed the post of superintendent of the United States Naval Academy on June 8, 2007. From the start of his tour, he was determined to take a clear-eyed and unsentimental look at the Academy's various customs, traditions, and practices and evaluate how closely they were aligned with the institution's stated mission and the Navy's priorities.

And Admiral Fowler was no great fan of the Herndon Climb.

Originally hailing from Bismarck, North Dakota, Fowler graduated from the Academy with the class of 1978. His career trajectory took him around the world through several high-profile assignments, including deputy commander of the U.S. Sixth Fleet. In the process he earned a bevy of medals and awards, including three Legion of Merit awards.

Before taking over as superintendent, Fowler's career rarely brought him back to the Academy. Sure, he managed to take in a few Army-Navy football games and attend an occasional Homecoming and class reunion; of course he's a lifetime member of the Alumni Association and he dutifully maintains a subscription to *Shipmates* magazine. But when the Navy tapped Fowler to head the Academy, they weren't getting a longtime Annapolis insider with sentimental attachments to traditions like the Climb.

"I had some different views on just what makes the Academy special," he explains. "I was focused on building leadership, to be ready to go to war in the Fleet. [The Academy] isn't college. This is different. And I just looked at all the things we were doing . . . and said, hey, do these things really align with what we're trying to do? Being a nation at war, being an example for the Navy?"

Another, much newer plebe-year tradition at the Academy was introduced in 1998 and has become very popular with midshipmen: the Sea Trials, a daylong series of physical challenges and contests where the plebes from each company are pitted against each other in something like the Olympic Games of combat readiness. The annual event is just the type of tradition that Fowler sought to emphasize and nurture in Academy culture: one in which the activity itself is closely associated with the Navy's objectives.

"It's important to get done with plebe year. That's a big event," Fowler says. "My goal was, and still is, that Sea Trials becomes the culmination of plebe year, reflecting a lot of things we want sailors and officers to do. Teamwork, overcoming difficult challenges, and more aligned with the Fleet. Everything I was trying to do was more aligned with the Fleet." Unlike the Sea Trials, in Fowler's view, the chaotic mass scaling of a greased-up Herndon Monument has little to do with preparing officers for combat or the unique challenges of shipboard life, or with commanding troops or following the military chain of command. The Climb, with all its attendant risks in the straining and tumbling and falling of unprotected bodies, seemed to Fowler like a tradition that was in sore need of unemotional analysis, especially in light of the Aggie Bonfire disaster.

"It's not like we don't want any contact," Fowler explains. "We have sports injuries all the time. People get disqualified; bad things happen. In a competition that is *regulated*, we take all the safety precautions we can. In soccer, football, wrestling, whatever it is. But here's something that's not regulated. There's no safety intervention possible. There's nothing. People climb this slippery thing and fall on top of each other, and it doesn't prove anything. It's not a competition with anybody. It's very entertaining but it's just not aligned with any operational risk management, any common sense. It's a tradition that [hasn't always been a tradition] so it's time to move on to something else. I thought Sea Trials was the right thing."

Should a tradition always remain in practice for no greater reason than that it's been around a long time and it provides the local civilian community with a free, two-hour show? Not necessarily, in Fowler's estimation. But despite his misgivings and his determination to avoid a Texas A&M–like tragedy on his watch he stopped short of ordering the flat-out cancellation of the Herndon Climb, which he felt strongly inclined to do in his first year as superintendent. He allowed the tradition to remain unchanged, and in 2008 and 2009 the Climb proceeded as usual.

"I said okay, fine, we'll do Herndon. It's very entertaining, all the cameras are there," he says. "But I'm also the person who looks at the moms in the crowd. And the moms write to me. And I look at everybody else during the two-and-a-half-hour entertainment event. Two-hundred, two-hundred-and-fifty-pound bodies falling eight or twelve feet on people's heads. Some of the other great traditions, like the Texas A&M bonfire, which got bigger and bigger and bigger, until it fell over and killed twelve students. Then they say, well maybe this isn't such a great idea. So, my thought was, I don't care how interesting or entertaining [the Herndon Climb] is, because it is very entertaining, except for the people getting fallen on and the moms worrying. So, I said, we're not going to do this anymore. I wanted to stop it and say, we're just doing Sea Trials."

Admiral Fowler had no desire to witness a tragedy happening in real time, before his eyes, especially if it was completely preventable and based upon a flawed premise that the risk of serious injury was worth the reward.

"I just didn't even watch it," he says. "I told the commandant to call me when the topper's getting close and I'll come over. I was not going to watch a person get paralyzed doing something I didn't approve of. But I knew this was going to happen. I mean, we're going to keep doing it until someone gets paralyzed, and then we'll say it's kind of a stupid thing we did."

The Climb may have had detractors among some Navy brass, many worrisome mothers, and students of vertebrate anatomy and emergency room doctors and nurses, who understand all too well the fragility of the human spinal cord and brainpan. But it also had its defenders somewhere in the Navy chain of command between Admiral Fowler and the commander-in-chief.

From somewhere in those dizzying heights, an order reached his desk. "Basically, I was ordered. No, you *have* to do it," Fowler recalls.

But as superintendent, he maintained a certain degree of latitude in how he chose to carry out the minute details of certain matters within his purview. And he was far from ignorant about the history of Herndon. He knew, for example, about the early days of the Snake Dance and when the first few classes of plebes managed to push a few of their number to the top of the monument. And he knew that in those bygone days, the monument was not greased.

So, if Admiral Fowler wished to improve the safety of the Climb, he could order that the monument not be greased. He wouldn't be cancelling the event or ending the tradition. Instead he was taking it back to an earlier time, arguably being *more* faithful to the history of Herndon. So that's exactly what he did. And the 2010 Climb, for the class of 2013, happened as scheduled, but there was no grease on the monument.

And when Keegan Albi scampered to the top of the clean obelisk on the shoulders of his classmates in an elapsed time of two minutes and five seconds, the joyful shouts of "Plebes no more!" were mixed with groans of disappointment from the spectators; and some midshipmen, including the plebes, were left with the nagging feeling that the challenge of Herndon had been made too easy. Albi and his classmates' achievement represent the second-fastest time ever recorded, matched only by Larry Fanning and the class of '72, who also confronted an ungreased monument.

Fowler wasn't surprised when the blowback against his decision, from Academy alumni and the community of Annapolis struck immediately and with hurricane force.

"It's good that we have opinionated, spirited alumni. I love them all. It's all great. [As superintendent] that's coming at you all the time," he says matter-of-factly. "The superintendent's job is like any command. If you're in the job to win a popularity contest, in command, you're in the wrong job."

During his tenure in the top post at the Academy, Fowler and his staff ordered a number of reforms, some of which improved his standing in the popularity contest and some of which did not.

He stepped aside in August of 2010, and his successor, Vice Adm. Michael Miller, ordered that the customary greasing of the monument be resumed. Admiral Miller downplayed Fowler's serious concerns about safety at Herndon and cited the importance of "conducting the ceremonial climb in the same manner as so many previous classes" to "instill spirit and camaraderie" and maintain a connection with the shared experience of previous classes.

Predictably, the class of 2014 needed a great deal more time to complete the task when they confronted a newly greased Herndon in 2011. Matthew Dalton ultimately reached the top after two hours and forty-one minutes, the longest time in fifteen years. Since then, no class has taken less than an hour and five minutes to cap the monument.

Pending future modifications to the current practices affecting the Herndon Climb, the completion time of each year's ritual will likely continue to be measured in hours and minutes, rather than minutes and seconds.

Michael Germano participated in the action as a plebe in the class of 2018. He wasn't a major player as his class succeeded in getting Javarri Beachum to the top of the monument to place the combo cover in a respectable time of one hour and thirty-eight minutes. But just a year later, Mike found himself appointed to one of the most thankless commands in the historical annals of the United States Navy: greasing the Herndon Monument in preparation for the Climb.

"There was a junior who was responsible for [supervising] the youngsters that was supposed to run the greasing process. He just sent out a random e-mail and my roommate and I were the first to respond. So naturally we were tasked with pretty much everything," Mike laughs, recalling this rare moment of naïveté from his career at the Academy, essentially volunteering for a distinctly unpopular assignment. "From getting the lard, to organizing the group, to getting up and actually putting the grease on. When you watch the Herndon Climb you never really think about it. Who put the lard on? How did they get it on? Like it just magically appears."

The task required Mike to lead a handpicked squad of similar "volunteers" on a critical mission deep into unknown territory: the cavernous kitchens and pantries of King Hall. So in the predawn hours well before first muster, the team kicked out of bed and stole their way along Bancroft's mazelike corridors, destined for the bowels of the mighty structure.

"They told us we would need to go get the lard from Services Division, a place we'd never been to. It took a while, we had to navigate our way through the depths of Bancroft that we never even thought about," Mike recalls. He and his classmates had a daily familiarity with the colossal dining area of King Hall, but not with the labyrinthine support facilities that served it. "Who even knew that it existed? Maybe it was underneath King Hall. Like the saying goes, you never really get to see how the sausage gets made. I never went back there again or even thought about it again."

This particular group of youngsters were on unfamiliar ground, but they soon made contact with those who labored in the windowless reaches of King Hall, where day was the same as night and there was a buzz of activity at all hours.

"They just wheeled out this massive cart, completely stacked with buckets of lard. It was kind of an unceremonious process, with a guy just handing over the massive *thing* of lard," Mike laughs. Next the squad somehow managed to maneuver this huge leviathan back through the kitchen, around the enormous rows of cooktops and bread ovens and stacked cases of foodstuffs, out of Bancroft Hall and onto the Yard, where a clean and unblemished Herndon Monument awaited, ready to receive its annual grease coating. It was still dark, as Mike remembers, and a fair amount of morning dew had collected on the granite.

"The process of getting the lard on there ranged from just smearing it on to kicking balls of it to just throwing it," he explains. "We go out there before dawn with a bunch of our friends, our company mates, and it was a lot of fun for us. We were out there playing, throwing lard everywhere. Throwing it on the [monument], throwing it on the Yard."

But Mike and the other youngsters were well aware that there was more to the task than just an opportunity to horse around with lard instead of snow in a mock snowball fight. They knew that they were obligated to complete the task of greasing the monument and do a good job of it, making the task of scaling Herndon as difficult as possible with the resources at hand. Their own predecessors had done likewise by them, and their competitive nature would not permit them to disappoint their classmates with a lackadaisical effort.

"No one wants to be the class that has the worst Herndon record at the Academy. You want to make sure [the plebes] don't beat you, that they're not going to have to bragging rights that come along with Herndon every year," Mike says. "Even the old guys that always come out every year. They say 'my class did this, my class did that.' The bragging rights never do go away."

Under Midshipman Third Class Germano's leadership, the 1st Company youngsters of the class of 2018 applied themselves to their assignment and completed it effectively, motivated by a sense of purpose despite their occasional diversion into horseplay. They used almost all the lard available, and as they regarded the results of their efforts, they were satisfied, despite the surplus, which they certainly didn't feel like lugging back to the pantry at King Hall.

"We ended up with one massive ball of lard, and we all just stood there looking at it," Mike recalls with a laugh. "My roommate [suggested], why don't we put it on top of the Dixie Cup? That will make it harder for them to get it."

The idea met with broad approval, although the task of getting a basketball-sized ball of lard to the top of the monument with no tools available other than a pair of extension ladders was itself an engineering challenge and teamwork exercise worthy of the Navy's best and brightest. Once they wrestled the ball into position, they finally judged their work satisfactory and immediately scrambled for their rooms—and particularly the showers—in Bancroft Hall.

Later, after they had cleaned up and changed into Herndon T-shirts and uniform camouflage pants, Mike and the other 1st Company youngsters were on the Yard again, engaged in a variety of tasks, from keeping overeager family members and tourists from mingling freely among the plebes, to making sure the plebes had removed their shoes and carried nothing dangerous or particularly useful into the Climb as they assembled along the walkway past Tecumseh Court. And once the cannon sounded and the great rush and crush around the monument began, they were right on the perimeter. And now they were armed with water hoses, and their spirit of competitiveness came into play once again.

"I don't know what other classes have done with the hoses, but we still didn't want 2019 to beat us, so we weaponized the hoses, for the most part," Mike admits, chuckling at the memory. "We were just trying to make [the Climb] a little bit more challenging. Like when someone got close to the top, getting closer, we concentrated fire on that plebe. Eventually we got yelled at and had to distribute [the water across them] to try to cool them down. Which, I believe, is what the hoses are actually intended for."

As for the huge lard ball at the top of the monument, it eventually slid down harmlessly into the chaotic meat grinder of plebes below. But for a while it became an object of concern.

"A reporter from *Navy Times* came by and said, 'Hey! Who's in charge here?' And we all sort of looked at each other, because nobody was really in charge," Mike laughs. "It was just 1st Company youngsters at that point. And I sort of like talking to reporters, so the reporter started asking questions and I was answering to the best of my ability. Until she said, 'Are you concerned at all about that massive ball of lard that could potentially fall and injure a plebe?' That's not something I'd thought about. I never considered that it could fall down and hurt someone."

Never mind that a number of human bodies had fallen from the pile since the start, each one much heavier and considerably less soft than a ball of lard; nevertheless, Mike stumbled over an answer to this particular question. And the answer he gave was not the one he would have given if he had the chance to do it again, nor was it what he expected his commanding officers would have preferred.

"I ended up saying something like, 'Uhh, if you get hit by the ball, you're just going to have to take one for the team.'"

"Obviously, it's a day that you look forward to as a plebe. So the excitement was overwhelming," recalls Chris Reaghard, USNA class of '95. "From what I remember, it was a bunch of mad midshipmen running toward this big, tall, all-of-a-sudden-it's-thickly-coated-in-white monument. The other thing I remember is, it was a really hot, sticky day."

He thought that the inch-thick coating was Crisco vegetable shortening, but whatever it was, it certainly didn't make the task of scaling the Herndon Monument any easier. Chris, a football lineman of appropriately sized proportions, settled in at the foundation of the human scaffold as it began to take shape. As for the sticky heat, he had no idea how long he and his classmates would be toiling in it.

"It took us a very long time," laughs Chris. "Eventually I got up onto the ledge, at the first level, locking arms with my fellow midshipmen. I remember there were multiple 'almosts.' But before the almosts, we realized that the Dixie Cup was duct taped to Herndon, so that was a not-so-pleasant surprise. It just seemed like there's no way we were going to get this thing off. Thank goodness we had Jim. He was able to pull himself all the way up and sit on top of that thing and peel that duct tape back one piece at a time. It was something I'll never forget, seeing him lifting that [Dixie Cup], and then putting the combo cover up in the air, and then placing it down. That was a great memory."

Before he became a plebe at the Academy, Jim Golladay was a Navy SEAL, earning his SEAL trident before accepting an appointment to the Naval Academy Preparatory School and ultimately embarking on an officer's career path.

Along that path he met Chris and classmates Shane Halloran, Dave Gordon, and Dan Morris. If those midshipmen had not decided to take matters into their own hands, the class of 1995 might *still* be out on the Yard, laboring to complete the Herndon Climb.

As a plebe, Jim was above average in age, one of the few who was over the legal drinking age.

"My friends and I were out having a very good time the night before," he laughs. "So the next morning at Herndon, I was not feeling great, and we were happy to stand on the outside and wait for everybody to finish what they were doing. Unfortunately, it just dragged on for hours and hours."

For Jim and his classmates, the time may have seemed considerably longer than it actually was, but whether it was one hour or five, they had seen enough. They were tired of it, and they simply resolved among themselves, "Let's go climb Herndon."

None of them was on the smaller side in the broad spectrum of midshipman body types. They were football players, all lineman-sized.

"I thought, we'll grab Dave Gordon and Shane Halloran, two big football players we knew from NAPS," Jim recalls, "and Dan Morris, another big, strong guy. And we'll just get our way in there, and we'll [improvise], and everything worked pretty much exactly how we drew it up."

Of course, they had drawn up no plan at all, other than to take Herndon by brute force alone.

"Dan got up, and I got on his shoulders. And I was tall enough to reach the top," Jim explains. There he found a layer of duct tape covering the Dixie Cup at the apex at least two or three feet in radius, according to his recollection. He dug his fingernails into the tape and held on, trying to free the cap, but with little success.

"Eventually things collapsed underneath me, and I fell down, and then I got the real, most memorable experience of Herndon," Jim says. "Being in the mosh pit. Everybody's just crammed in there, and it's about 150 degrees and people are getting trampled on and screaming, and they think they're dying. That's a real wake-up," he laughs.

Shortly, Jim and his mates made their way to the sidelines again, where they watched for a while as their classmates continued to struggle, with futility but no lack of fortitude, against Herndon's slippery surface. But before long, Jim's well-developed leadership instincts, incompatible with avoiding the action, fired his ambition to finish the Climb.

"I told them, obviously something drastic is going to have to happen, because I was literally hanging on it, and it's not coming off. So, let's try again," he recalls. This time, he knew that he had to find a way to get a little bit higher, to somehow get an elbow up over the monument, to give him the proper leverage to work the Dixie Cup free from under the layer of tape. So he pulled Dan, Shane, Dave, and Chris together again and rallied them to his side, and this collective half-ton of midshipman muscle and blood and bone threw itself into the action a second time.

"We went in and basically did the same thing we did before," Jim explains. "This time I got up to the same point and could not get any farther. And Dan Morris got his hands underneath my feet and military pressed me up that extra foot and a half, where I could reach over the top. And I just pulled myself up and sat on top of Herndon."

"I don't believe I'm strong enough to actually press him up," Dan says modestly. "I imagine he was pulling himself up, to some extent, at the same time. But we eventually pushed him up, and he ended up sitting on top of the monument, working away at all the duct tape and finally getting the Dixie Cup off, and then putting our cap on. Dave Gordon, Chris Reaghard, and Shane Halloran, those were the three big guys that I basically stood on. They propped me up, pushed me against the [monument]. Jim climbed me like I was a ladder."

Dan's recollection of the night before Herndon is considerably more vivid than his friend's, and he also places a great deal of faith in the notion that the statute of limitations has long expired on any plebe-year rule violations.

"We went out the night before, I think without authorized liberty. We just took our own liberty," he laughs. "We got *wasted*. I mean, we were probably still drunk when we woke up the next morning for reveille and had to go do Herndon. Neither one of us had planned to participate in Herndon at all. So when the whole class sprinted down Stribling Walk and ran to the monument, Jim and I walked over to a tree and just sat down in the shade, drinking water and Gatorade and just trying to recover. Herndon was our last military obligation of the year, and once we finished it, we could leave."

What a relief that would be, not only for Jim and Dan and their friends, but for all the plebes of the class of '95. Although the final recorded time of their successful scaling of the Herndon Monument would be two hours and twenty-one minutes, in the moment it seemed more like three or four hours to Dan, who was particularly eager to get out of town and who might have been overly optimistic about meeting his planned travel timetable.

"The funniest part of the whole thing is not necessarily that Jim and I just decided to go do this, but the reason was that I had a train to catch," Dan says between chuckles. "I said, 'Jim, if we don't get the [expletive] hat down, we're not getting out of here!' "

Shane recalls that some of his classmates were less than pleased when they suddenly found themselves bearing the weight of nearly an entire offensive line of football players.

"Normally, you don't get the bigger guys going for the next tier," he explains. "[Dan Morris] gets up on my shoulders, and he and I are about the same size. I was 227 back then, he was probably about 230. A couple of big guys like that crawling over people, you get some really angry looks from the smaller classmates. Hey man, you are too big. I literally got cussed out: you are *too big* to be climbing on me! I thought, hey dude, you need to go eat a couple more sandwiches. You're coming to Herndon, bring your game!"

Dave Gordon is unlikely to forget that amazing day, the last of his plebe year. In fact, he thinks of it every time he has a flare-up of shoulder pain.

"I had Dan Morris' foot on my shoulder, as he's standing there, pressing Golladay," he laughs. "I give him a hard time whenever I speak to him. I tell him the reason I have such bad shoulders now is because of him standing on me."

When Jim reached the top of the monument for the second time, after extra boost from his classmates, pulling himself up to a seated position at the apex, he was able to more effectively attack the layer of duct tape that had so bothered the other climbers who got near the top but fell down again, unable to remove the Dixie Cup.

"I started stripping the tape off, wedging my fingers under and stripping it off. Then I'd rotate ninety degrees and do the same thing until I pulled the cover out from under me," Jim explains, before invoking a metaphor that betrays a keen interest in Greek mythology. "It looked like the Gorgon's head as I was holding it up, with three feet of duct tape hanging off the thing."

His classmates passed a combination cover up to him. And he placed it on the top of the monument with a dramatic flair before sliding down, unceremoniously and painfully. "And I gave myself two good slices across my biceps, where the corners of Herndon were unforgiving," Jim recalls.

"It was high fives all around, and we were all pretty excited," says Dave of the Climb's muscular conclusion. "It was definitely memorable, how it all went down. Knowing that you have a foot buried in your shoulder, and if you weren't going to make it, if you weren't going to be able to sustain that much weight and pressure, then the whole thing was going to tumble. As I look back on it, to me it is definitely one of those things that epitomizes what they try to teach us at the Academy in terms of leadership and teamwork. When it comes down to it, a small group of people with the right type of purpose, motivation, and direction can get a lot accomplished. This is what we want our junior officers to do."

The result of all that effort was a successfully completed Herndon Climb. The final time wasn't one for the record books, but under the circumstances, the class of '95 had no reason to hang their heads. On the contrary, they did just the opposite, celebrating their liberation from a year of sustained challenge.

And Dan Morris, his steps buoyed by the knowledge that he had ascended from plebe status, made it to the train station with time to spare.

"Before Herndon, you're a plebe. You have no rights, no privileges. All of your American-given rights are taken away from you. Albeit voluntarily," explains Jason Cronin, USNA class of 1987, from the perspective of a seasoned Navy officer, a helicopter pilot and twenty-year Navy veteran who served in both Operation Desert Storm and the 2003 invasion of Iraq. "You gave them up, at times you wonder for God knows what. But after Herndon you're no longer a plebe. You're no longer a thing; you're a person, you're yourself. And you also become part of something bigger."

For Jason, the Climb itself certainly made him a part of something bigger, in a most unexpected way, in the crush at Herndon.

"Basically, you've got to have a strong foundation if you want to get four layers of people up there. We were packed so tightly that I literally could have picked my feet up and I wouldn't fall," he recalls. "You get to a certain height and depending on what you need—oh you need a solid guy, or you need a tall guy—people were getting lifted up and just walking across the top of the crowd pressed to the monument. I remember a couple of guys getting lifted out and encouraged to walk across our shoulders, and heads. People were putting their hands up. It just didn't matter, let's get it done."

Decades later, so many years removed from his days at the Academy, Jason remembers feeling the sense of nostalgia that was clearly invoked by the Climb, visible in the faces and body language of what he thought of, at the time, as the "old, old timers."

"They were standing and cheering. [We saw] them on the steps of the Chapel . . . the glint in their eyes and the smiles," he says fondly, now that he has reached an age where the modern midshipman might consider him an old-timer. "It's not about wanting their time to be better than ours, but you can almost see them reliving their youth. The funny part is, I remember thinking to myself, these guys are so old, they're looking at us and wishing they were us."

An Annapolis resident, Jason participates in the Academy's sponsorship program, offering midshipmen an occasional opportunity for a home-cooked meal, a nice place to spend the night away from Bancroft, and the sympathetic ears of a surrogate family. So he interacts frequently with them, and although he seldom attends a Climb, he is always well aware of it.

"My young daughters and my wife, they go out and watch Herndon, and they come back with pictures, and I see all these old guys still watching, and still with that smile," Jason laughs. "I started realizing: You know what? I remember in the midst, thinking here are these old-timers, wishing they were me. And I am one, now."

13

A FINE TIME FOR A CLIMB

(or, One Hot Monday in May)

I long to accomplish a great and noble task, but it is my chief duty to accomplish small tasks as if they were great and noble.

—Helen Keller

Sunday, the nineteenth of May 2019, was a particularly warm day in Annapolis. The mercury struck a high of eighty-three degrees at midday, with a relative humidity of 66 percent under sunny skies, occasionally broken by scattered low clouds, satisfactorily muggy conditions to cause any normal human body to break into a sweat with minimal effort.

But the gods of weather around Chesapeake Bay are a mischievous and sadistically inclined lot. Somehow they knew that the annual Herndon Climb was scheduled for the following day, and they were not content to allow a continuation of the present conditions, too mild to really test the fortitude of the Naval Academy's latest class of plebes, this time the group that would graduate in 2022. So they cranked up the temperature by a notch, settling on a nice round figure of ninety degrees Fahrenheit, and they added another bit of moisture to the warm atmosphere, increasing the relative humidity factor by another 4 percent.

Not until the following day would the oppressive heat and humidity abate. On Tuesday, as the acclaimed Blue Angels stunt pilots roared through the skies over Annapolis in their F-18 Hornet fighter jets, buzzing the wide-eyed and appreciative masses on the ground, temperatures would drop to a much more pleasant seventy-four degrees and the humidity would ease a bit, to 64 percent. By then, the Climb would be complete, the fourth-class midshipmen would be "Plebes no more," and the only people complaining about the previous day's weather would be the spectators.

Midn. 2C Jessie Baynes had a job to do, and like any good midshipman, she would make absolutely certain that the job would be done. She had taken an oath, and she was duty-bound, even at the cost of her life, to defend the United States Constitution and its principles with every ounce of strength in her body and every resource within her command.

But getting covered in lard? That's where Jessie drew the line.

The job assigned to her, supervising a squad of 1st Company youngsters in applying a coating of lard upon the Herndon Monument to prepare it for the Climb, seemed like a fairly mundane task. With more than a dozen able-and-willing troops available and a year of seniority over them, Jessie exercised the privilege of leadership and elected not to participate directly in the greasy, hands-on work. After all, immediately after completion of the current assignment she was scheduled to report for a fitting of her Marine Corps gear, which was necessary for her upcoming summer training. It just wouldn't do to be liberally smeared with pig fat when she showed up for the fitting.

They had sufficient manpower for the job, but the only tools available to the youngsters were a pair of extension ladders and their own hands. A hundred pounds of lard, in two fifty-pound boxes, was requisitioned from the King Hall pantry and now sat on the grass, gradually softening under the warmth of the rising sun, ready to be applied to the Herndon Monument as part of the tradition that reaches back to the 1940s.

Handling the lard was turning out to be problematic. The boxes had been kept under refrigeration, and now the youngsters discovered the downside of that approach, since lard and other shortenings, like butter, tend to harden at low temperatures.

"This stuff is too hard!" one youngster complained, pressing a finger into the fatty bolus. It was workable, but hardly *spread*able, approximating the consistency of modeling clay or cold Play-doh. "How are we supposed to handle it?"

"With your hands!" replied a sarcastic classmate.

A couple of bricklayer's trowels would have been useful, and maybe a blowtorch. But lacking those implements, the youngsters set to work.

One year earlier, on the morning of a similarly warm Herndon day, the youngsters of the class of 2020 faced a different challenge. Prior to the monument preparation, someone made sure that a single, hundred-pound container of lard, earmarked for Herndon, would be available at first muster on the day of the Climb. The purchase order for one bucket, however, was placed through a different food wholesaler than the one that had previously been used, and the new vendor sold that particular commodity only in thirty-seven-pound buckets. So with barely a third of the planned quantity, there was not enough lard to do justice to the task of applying a thick coating to the entire twenty-one-foot obelisk.

Undaunted, a detachment of two youngsters from the squad volunteered to make another raid on the King Hall kitchen and came away with nearly sixty pounds of salted grade AA butter in one-pound blocks, each block wrapped in waxed paper. For good measure, they begged or borrowed or outright stole two gallon-sized jugs of olive oil and two extra-large cans of Pam cooking spray. The butter was an imperfect substitute for lard or vegetable shortening, which turns to liquid at a much higher temperature, but in a pinch it would have to be good enough. And once it was liberally applied, mixed in places with lard-covered sections, the Herndon Monument resembled nothing so much as a colossal, well-buttered breadstick. The Pam cooking spray was not used, however, when it was applied to a small area and proved ineffective as a lubricant and potentially damaging to the time-tested surface.

And when the youngster bravely standing on a greasy extension ladder poured the two gallons of olive oil onto the very apex of the monument—just prior to placing the all-important Dixie Cup—dozens of glistening legs of golden oil trickled slowly down the four sides of the monument in gently winding streams that cut tiny channels through the buttery, lardy surface in a display that was not only visually pleasing, but downright appetizing.

"Hey, that looks good!" an early-arriving spectator razzed the youngsters from the sideline. It was all in fun, of course. "Can I have some for my English muffin?"

Later, the improvised recipe began to seem sadly inadequate as the morning wore on and the mercury rose higher. By the time the Climb commenced, most of the butter and oil had melted away, leaving a monument that seemed less well-lubricated than in years past. But such concerns were alleviated when the plebes of the class of 2021 needed well over two hours to lift Peter Rossi to the top, where he had the honor of placing the precious combo cover.

This year, the youngsters had plenty of lard available. All they lacked was a convenient means of applying it to the Herndon Monument without getting their hands thoroughly greasy, and its cold hardness would necessitate warming it, kneading it slowly by hand into a spreadable compound, a process that was certain to increase the messiness of the task.

Meanwhile, pallets loaded with several hundred bottles of water arrived on the scene. These were intended for the climbers, the plebes who would later be laboring and sweating in the hot sun for as long as it took for them to get one of their number to the top of the monument. But now those bottles needed to be moved, by the caseload, from the stacked pallets to a convenient location within the bounds of the temporarily fenced-off grassy area near the monument. Many of the youngsters seemed more willing to carry water bottles than to bury their arms in lard, so the unloading process was speedily completed.

Eventually, everyone pitched in, embracing this unusual assignment with typical midshipman humor and jocularity and the ever-present willingness to work hard and contribute to the squad's success.

Everyone, that is, except the fastidious Jessie, who remained in more of an observer's role, chatting with interested spectators some distance from where lumps and clumps of lard flew and splattered across a wide radius around Herndon Monument. She was approached by one of the youngsters from her company who had applied himself so enthusiastically to the assigned task that he now appeared to be wearing a full-body coverall of grease; the yellowish-whitish goo coated his hands, arms, blue-rim T-shirt, and Navy athletic shorts. Globs of the substance shone in his hair and clung to his ears. Jessie politely demurred when the youngster jokingly offered her a full-body embrace, but she accepted a brief handshake, and then immediately regretted that decision and started looking around for a towel. There was none to be found.

Performing more of the commanding officer's duties on the scene was Joe McGraw, class of 2020, now a second-class midshipman at the tail end of his third academic year.

"It's really tremendous, this long line of tradition, going back to the 1940s, larding up the monument," Joe observes. "It symbolizes the end of plebe year, a long hard year." Later he would have the honor of speaking about the great history of Herndon when he took the microphone and addressed the crowd of spectators and the plebes before introducing Superintendent Carter for the official start of the Climb.

As a plebe in 2017, Joe was persistent and lucky enough to succeed in placing the combo cover on the top of Herndon for his class in an elapsed time of two hours, twenty-one minutes. "It was surreal. Absolutely the most surreal experience of my life," he recalls of his success in the Climb and the joyous moments that followed it.

The following year, as a 1st Company youngster, he participated in the greasing of the monument with the memorably savory mixture of butter, lard, and olive oil. Today, he found himself nominally in charge of the job, ordering the more recalcitrant youngsters to participate, directing the ladder-raising, instructing everyone to use the warmth of their hands to work the stubbornly resistant lard into a smoother and softer consistency, which made for much easier application.

The learning curve was short for the high-achieving youngsters, and soon they were scooping softball-sized lumps from the bucket, rubbing them with their hands, and pressing them against Herndon's granite surface. Then the lumps were gradually spread out to an even thickness, each one covering an area of roughly two square feet. The work proceeded slowly but steadily, and entirely by hand.

One youngster devoted his efforts solely to scooping out portions of lard, like an expert maker of snowballs preparing for a lengthy siege on a snow fort. He worked the lumps into smooth, presoftened balls and stacked them in a convenient pile for use by his classmates, who handled the spreading.

Soon the youngsters had large sections of the lower reaches of the obelisk well-covered, and the extension ladders clattered to full length. Three or four classmates lifted the ladders with greasy hands and maneuvered them into position, one on each side of the monument. One brave volunteer scaled each ladder, and now the squad's year-long practice in teamwork became apparent as lard balls were tossed upward and—mostly—deftly caught and applied to the smooth granite.

Once the ladders were moved to the other two sides of the monument and the same tossing and catching and smooshing and spreading was accomplished, the upper and lower thirds of the monument were well coated in lard. But the middle third presented a more vexing problem: it was out of the youngsters' reach while they stood on the ground, and to reach it from a ladder, that ladder would need to be positioned at an angle that was either too acute or too oblique to climb safely.

Now the fat *really* began to fly as the youngsters improvised a solution. They tested the strength of their throwing arms and rotator cuffs, hurling fastballs of lard against the granite, most of which stuck fast and some of which slipped off due to their being either not softened enough or not thrown hard enough. In all cases, the back-splatter was considerable, and showers of grease droplets rained down on all the participants and those observers who dared to get too close to the action.

One errant grease bomb flew quite high and wide of its mark, missing the surface of the obelisk entirely and sailing past the boundary of the temporary fence, landing in the back of a flatbed maintenance truck whose driver did not see any of the humor that the youngsters saw in the situation. He unsmilingly demanded that the gooey mess be promptly removed from his vehicle.

When the two boxes of lard were finally empty, McGraw and the others stood back to admire their handiwork, satisfied that the monument was well and thoroughly greased. It had a unique appearance, evenly coated on the upper and lower sections, pimpled with lumps in the middle. They etched a few messages in the lard coating, such as "plebe year never ends" and other words of dubious motivational value. The grass around the monument was slick with drips and spatters of lard as they posed for a group photo.

Before dismissal, McGraw rebriefed them about their duties during the Climb itself, which would begin in about six hours. Then, with greasy hands, they gingerly picked up their cell phones from where they had been placed nearby, outside of the danger zone, and headed back to their rooms in Bancroft Hall to get cleaned up and to change their clothes.

And Jessie Baynes, relieved that the greasy task was finally completed, headed off to be fitted for her Marine Corps gear.

If midshipmen Welch, Oweimrin, and Nygard had any apprehension about the upcoming Herndon Climb, in which they and their classmates would be expected to match the success of their predecessors in the decades-old tradition, it didn't show in their attitudes or their appetites as they devoured pepperoni pizzas in the Drydock Restaurant at the northwest end of Dahlgren Hall.

The Climb was scheduled to start in just ninety minutes. But the three plebes from 28th Company seemed nonchalant, chatting amiably among civilians and retired Navy officers alike. They brushed off their poor showing in the Sea Trials, where they finished twenty-ninth out of thirty companies, pointing out that their company had been significantly depleted during the year by attrition and by a spate of injuries and personal medical issues that reduced their fighting strength to the point of having much smaller squads competing in the various events. This meant that they had no opportunity for rest, and many of them were compelled to face certain competitions twice, such as the famous pugil-stick battles. And after all, another company scored even worse than the 28th.

Now, having survived a typically brutal plebe year and a humbling Sea Trials performance, the prospect of a unique challenge like the Herndon Climb didn't worry them at all.

"We'll break the record this time," Wesley Nygard predicted, perhaps only half in jest. "Thirty minutes flat."

"We've been training all year for it," Jadd Oweimrin said with a laugh, but his comment contained a grain of truth. Although it is a long-standing Academy custom that plebes are not allowed near the Herndon Monument all year, these three men had in fact been training all year, maturing and advancing, physically and mentally, and they all had successfully conquered challenges both great and small. Now, quite simply, they were ready for the next challenge, which happened to be the Climb.

Not a single droplet of pizza sauce marred their pristine white summer uniforms, so if their appearances were any indication, the future of the Navy will be in good hands when young men like these expect to be commissioned as officers with the class of 2022.

How would they describe their plebe year at the Academy?

"Long."

"Long," each midshipman answered in unison, between bites of pizza.

When pressed as to whether a particularly well-known, profane adjective that has been in common usage on Navy vessels for centuries could be accurately applied to their retrospections of the academic year, they also concurred.

No rights whatsoever? Considered the scum of the Earth?

"You could say that," Brogan Welch answered resignedly.

"They pretty much just strip everything away from you," Wesley added.

All that would change in just a few hours. Once their class completed the Herndon Climb, they would gain a host of new privileges, revel in a newly earned higher status, and proudly shed the burden of being plebes.

Midshipmen Welch, Nygard, and Oweimrin didn't know much about the Herndon Climb beyond the facts on which they had been thoroughly briefed. They knew the objective: to work together, to remove a Dixie Cup from the apex of the monument and replace it with an officer's combination cover. The emphasis on safety and the importance of hydration were well in mind.

But they knew little of the rich history of the Climb, and they were unfamiliar with Herndon's horror stories, like the four-hour ordeal of the class of 1998, and of Steve Fortmann's heroics in finally bringing it to a close. As they listened to that epic story, their own confidence that they might set an all-time record for speed became mixed with apprehension.

They knew of the notorious reputation of some of their predecessors at the Academy following the 1976 gender integration, which happened long before they were born. They knew that female midshipmen have been cruelly excluded from participating in the Climb, or pulled down while attempting to scale the pile of bodies. In past decades such a dishonorable practice was more common and more widely tolerated, and by some accounts it continues in a few instances.

They were well aware that if the "scaler" for their own class happened to be a woman, that would represent a historic first. Their enlightened views of such a prospect are evidence of the ongoing sea change in the treatment of women at the Academy over the course of nearly five decades; to a man, they supported the idea of a female classmate topping the monument, and they pledged to offer assistance, not resistance, to their efforts at Herndon.

"If she gets up there first, I'm more than happy [to help]," Brogan promised.

"I'll lift whoever climbs up on top of me," said Wesley.

But making history, either by having the slowest or fastest recorded time or by breaking the gender barrier, was far from their minds. What mattered was simply completing the task. They would have to wait until Graduation Day, at the end of the week, to add a diagonal stripe to their uniform sleeves and shoulder boards, announcing their official ascension to the rank of midshipman third class, or youngsters. But according to the widely observed Academy custom, completing the Herndon Climb would grant them wonderful new privileges and respect, for a few days, as midshipmen fourth class.

It has been said that no slice of pizza has ever gone uneaten when there are college students or active duty service members present. Brogan Welch, Wesley Nygard, and Jadd Oweimrin, after finishing their last meal as plebes, scrambled back to duty, with just a few minutes to change out of their crisp white uniforms and into Navy athletic shorts and T-shirts and join the rest of their company mates.

They had a date with Herndon.

The afternoon sun shone hot and bright as the Yard filled with spectators of all ages eager to watch the 2019 Herndon Climb for the class of 2022. Some of them spread blankets on the grass or lounged in lawn chairs, listening to the music of the Academy's own jazz and swing band, Trident Brass, who were set up on the bandstand. The band worked through a fine selection of contemporary and classic American tunes, ranging from a soaring "Skyfall" to the ever-catchy "Mack the Knife" and Cab Calloway's racy Depression-era classic "Minnie the Moocher," with a clutch of Sinatra standards thrown in for good measure.

Meanwhile, food stands and catering trucks did a brisk business in funnel cakes, grilled sausages, and Italian ice. Cold drinking water was at a premium, and any shady spots around Herndon were crowded with people, shade from the hot sun being a slightly higher priority to many of the spectators than an unobstructed, close-up view of the monument, where the drama and spectacle of the Climb would soon unfold.

Neither brutal heat, nor sticky humidity, nor the prospect of an incredibly difficult physical challenge could defeat the enthusiasm of the brigade's plebe class as they waited on the red-brick walkway between Tecumseh Court and the Herndon Monument, corralled by the low temporary fence of blue plastic mesh. They knew that in a few hours at most, barring some sort of unforeseen event that had never happened in the history of the Climb, they would be able to celebrate and officially throw off the burdens of plebe year.

At the head of the line were the members of the Iron Company, the plebes of the 14th, whose triumphant performance in the Sea Trials just a few days earlier, outscoring twenty-nine other companies, earned them the honor of leading the assault on Herndon.

Their eagerness to seize that honor and throw themselves into the challenge was evident as they sang and chanted and engaged in play fighting and good-natured roughhousing. Giddy with excitement, they laughed and posed for pictures, some chatting and bantering with friends and family on the other side of the loose plastic fence. A few plebes who evidently had elected to stay out of the roughest action around Herndon brought their cell phone cameras to the event; they were obligated to make a solemn promise to share the results with those who posed in small, chummy groups, mugging and preening, always with their arms around each other's shoulders, or showing off their flexed biceps and tightly wired torsos for posterity.

This palpable exuberance, almost like that of children waiting on line for a roller coaster ride, was not confined to the 14th; it was on full display down the assembled line of companies. All the plebes were barefoot, since socks were now officially banned from the Climb for the first time. Many of the men had ripped open their blue-rim T-shirts, intending to peel them off completely once they were within throwing range of the greased monument.

Many had fashioned warrior-style headbands and armbands from strips of black cloth or pieces of T-shirts. All that was missing was war paint for their faces.

Confident to the point of cockiness, they planned, or perhaps fantasized, about strategies.

"Once we get close, they'll already be stacking up," one plebe mused aloud. He and his company were near the end of the line, courtesy of a less-than-stellar performance in the Sea Trials but showing no outward signs of humility or disgrace as a result. "So all I have to do is look for a good place to climb up. Bang! It's done, and we're outta here!"

"So, you know it's covered in some kind of grease, right?" a classmate inquired.

"Pig lard," a nearby civilian offered helpfully. "It's covered in a hundred pounds of pig lard."

"Sounds good to me!" laughed the confident plebe, shaking his head vigorously and wagging his extended tongue.

"You're disgusting!" observed his classmate. Naturally, the rapprochement brought on an even more exaggerated display of tongue wagging.

All this unrestrained misbehavior on the part of the plebes was good-naturedly tolerated by the upperclassmen who were charged with maintaining some degree of order. Most of them wore camouflage pants, combat boots, and T-shirts emblazoned with the words "Herndon Staff." In addition to general crowd-control duties, they were tasked with strict enforcement of the extensive safety protocols developed for Herndon. Once the Climb was under way, they would also be responsible for distribution of water, in bottles for drinking and also delivered via hoses for the purpose of cooling the pile of straining, overheating bodies.

The upperclassmen, all second and third class, understood the plebes' current state of gleeful, boastful rowdiness. They had been there themselves.

And next year at this time, the current class of plebes would be on the other side of the fence. Some of them would be assigned to staff duties, and others would be there just to watch the action, wearing summer white uniforms and mingling freely among civilian onlookers, happy with the privileges of upper-class status. Maybe they hoped to relive some of the excitement they had experienced when they participated in the same event.

Next year, they would all have a new batch of plebes to put through the training wringer, making their lives miserable and more difficult every day of the week. But at the end of the year, they would all be cheering for those plebes, at Herndon.

Hidden behind a pair of midnight-black aviator sunglasses, the eyes of Vice Adm. Walter Edward "Ted" Carter, sixty-second superintendent of the United States Naval Academy, observed the chaotic scene as it played out from his vantage point at the Zimmerman Bandstand, where the Trident Brass band sat ready to tear into a spirited version of "Anchors Aweigh" once he finished his remarks and signaled the commencement of the Herndon Climb.

"Class of 2022! Are you ready?" Carter addressed not only the overeager plebes, roughly twelve hundred strong and impatiently vibrating in place, but also the ever-growing number of spectators, who crowded into the best viewing spots, with shady places beneath the leafy trees being the most desirable.

If Carter's resolute posture and the broad smile that crossed his features were any indication, those unseen eyes were fairly beaming with no small measure of pride and satisfaction as he presided over the fifth Climb of his five-year term as superintendent.

"On the count of three, I'm going to do a 'Damn the torpedoes,' and you say, 'Full speed ahead!' and that's when you're ready to go," Admiral Carter announced. "One! Two! . . . Three! . . . Damn the torpedoes!"

The anticipated reply of "full speed ahead" was drowned out by a generalized roar of approval from the assembled audience, accompanied by the plebes of the class of 2022 themselves and the familiar strains of the Navy fight song, delivered in a spirited rendition by the horns and other instruments of the Trident Brass.

Having fulfilled his obligation to preside over the opening formalities and with the Climb fully under way, Carter walked across Blake Road and ascended the steps of the Chapel, where a makeshift reviewing stand overlooked the action and the Chapel's magnificent dome provided welcome shade from the blazing sun. Today the dome was covered by scaffolding to accommodate an extensive cleaning and restoration project, creating an interesting juxtaposition with the human scaffold being hastily erected around the Herndon Monument below, built from the bodies of plebes.

Relaxed and circumspect, accompanied by his wife, Lynda, Admiral Carter chatted amiably with friends and visitors, active duty and retired Navy officers, and a handful of lucky civilians. His pride was evident as he discussed the long history of the Climb, the noble tradition it represents, and its beloved status at the Academy and among the townspeople of Annapolis.

Carter's next official duty at the Climb, his fifth as superintendent, would be to personally congratulate the eventual scaler of the monument and present that individual with a pair of admiral's ceremonial shoulder boards, perpetuating the supposed legend that the Herndon scaler would be the first member of the class to reach flag rank. He shared his opinion about which of the recent scalers most decidedly caught his attention and might one day fulfill the promise of the legend.

"Chris Bianchi," said Carter, naming the midshipman who topped the monument for the class of 2019 and was just days away from commissioning. "I just have a feeling about Chris. I think he might be the one."

Time will bear out whether Carter's prediction proves to be correct. And if a knowing twinkle of prescience lit up his eyes, no one could tell, because they were still concealed behind those dark aviator sunglasses.

The Herndon Climb started with a bang—literally—as the cannoneers of 5th Company filled their traditional role, firing a noisy twelve-gauge blank shotgun shell from their miniature 18-inch cannon. The loud report would be repeated every thirty minutes until the Climb was finished, and it was always startling, and not only to children and other civilians among the spectators.

"Somehow, you always think you'll get used to the [cannon blast], but you never really do," observed an Annapolis local resident and Herndon devotee.

The Iron Company led the charge, and on their heels came the rest of the eager plebes, rushing forward and surrounding the Herndon Monument. Blue rim T-shirts filled the air, pelting the well-greased obelisk, to be used by the forward troops as rags to wipe away as much of the lard as possible from the slick surface. Within a few minutes, construction of the rudimentary human scaffold had begun.

Progress seemed to be extremely rapid, and the excitement of the crowd reached a fever pitch. Spectators wondered, would this year's Climb be short and sweet, the equivalent of a first-round knockout in a boxing match?

The answer was no. Herndon would not be conquered so easily, as nearly every other class in the past half-century has learned.

Somewhere in the rickety, hastily built stack, something gave out. A foot slipped from a shoulder, or there was still too much room between adjacent ground-level bodies to prevent the slightest wobble from throwing off the balance of the delicately perched climbers, or someone shifted their weight at just the wrong moment, or any one of a thousand other tiny things happened and was amplified into a disaster, bringing everyone crashing down in a seismic collapse.

The first major spill of bodies from the face of the monument brought audible gasps, even cries of despair from the assembled onlookers. Veteran observers know that these collapses always happen, bodies will fall from significant heights throughout the Climb until it is finished, but parents attending the event for the first time are stunned and aghast at the idea of their son or daughter being somewhere in the meat grinder around Herndon.

If those worried Navy moms and dads could have spoken to Fifth Battalion Cdr. Kelly Laing, some of their concerns might have been alleviated.

"This whole packet is the operations brief," said Commander Laing, observing the action from the superintendent's reviewing stand on the steps of the Chapel and brandishing a multipage stack of densely printed 8 × 11 sheets of paper. "The safety concerns, from how we brief the midshipmen [plebes] on the Friday beforehand. How to make sure that the kids have tight basics, and if somebody starts to fall, reach out and catch them. It's not that you can blindly stand at the base, you actually have to watch what's going on. We're absolutely clear about that."

According to Laing, safety is the highest priority for the officers tasked with organizing the Climb, and the protocols are regularly reviewed and updated. "When it's over, I'm going to sit down and debrief everybody on the lessons learned and what went right, what went wrong, what we can do better next year," he says. "This year we have video and different media set up, so we've changed the layout of the land a little bit."

In recent years, very close attention has been paid to the participation of women in the Climb. Part of the pre-event briefing includes a stern warning to anyone seen deliberately pulling a female classmate down.

"I know that the class has been briefed very solidly on this. Hey, no foul play on this one, all right? Help each other out. Male, female, doesn't matter," Commander Laing points out. "Knowing the midshipmen for the time I've been here, from communication in Bancroft Hall, I actually think that they *want* it, they want the female midshipmen to succeed. I really do. When we take from a cross-section of society, are we going to remove everybody's prejudice and bias, or what they came to the Academy with? By the time you get through a four-year institution? No. But I think we can root out most of it, and I believe we have. I really do."

Heat exhaustion, which can have very serious repercussions, continues to be of paramount concern from a safety perspective, and the protocols thoroughly address the means of alleviation and the importance of immediate medical care for any victims.

"The biggest safety priority is keeping the midshipmen cool. Obviously, the water helps with that," says Laing, pointing out the ever-present hoses, which some observers assume are intended to make the task more difficult for the climbers. "This year, we actually have more people here than we did last year. We've got a staff of foremen. I have about thirteen foremen that are watching and knowing the exact protocols for when to pull somebody out."

He agrees that the Climb has its inherent risks. But he disagrees with those who would attempt to improve safety to the point of changing the game entirely, by disallowing the application of grease to the monument.

"Inevitably, there are going to be strains and minor cuts and scrapes and bruises. We understand that," Laing points out. "But we also don't want to develop a fighting force that we're unable to send out into the desert, send to the front lines, because we pampered them, treated them like pansies the entire time they're here. We've got to have people that are used to doing this kind of thing. [The Climb] is that same kind of teamwork building, even though it isn't combat. It's about the mental fortitude, accepting challenges and getting out of your comfort zone."

The chaos and the sweaty struggle of a Herndon Climb certainly seem well outside of anyone's comfort zone. Despite his position as battalion commander and his key role in coordinating the event, Laing never participated in a Climb and did not attend the Academy himself. But his admiration and respect for the Academy's customs and traditions—particularly this one—are apparent in his dedication to the task.

"I'm a huge believer in this, in the institution and everything that it represents. The raw struggle, the teamwork that it takes, the culmination of a year of full-on effort. It's absolutely incredible," he says. "Last year, I arrived a week before the Herndon Climb. That was really my big introduction to the Naval Academy, standing on these exact steps [of the Chapel] and watching Herndon, and just seeing that struggle and talking to the midshipmen afterwards about what it meant to them."

"It was complete chaos," said Joe McGraw, recalling his own experience in the Climb as he watched the current one in progress. "But that's how life is, you've got to take it by the horns and charge in there. Some people try to come up with plans, but it all, always, just falls apart. It never survives first contact with the enemy. They'll get up there eventually. Probably, hopefully after not too long. It's pretty hot out here."

Whatever plans the class of '22 plebes had come up with were quickly discarded in favor of the time-tested approach that generally results in success, although that success might not always come immediately.

They continued sweating and straining in the blazing midday sun as a second loud blast from the cannon signaled that thirty minutes had elapsed since the beginning of the Climb. No new speed record would be set today.

Slowly and methodically the plebes began working together, and soon they began to understand the potential of a solidly engineered structure built from their own bodies. They packed themselves together tightly, forming a strong ring around the base of the monument, duplicating this approach for a second layer, and always pushing, pulling each other up. Inevitably, though, came the disappointments, the partial collapses that sent some of the climbers tumbling down, as weary limbs gave out and the slippery mixture of grease and sweat and water—with a generous portion of mud—worked against their success.

Climbers became supporters as some plebes stood on the shoulders of their classmates and had others stand on theirs. Higher up on the monument now, in greater numbers and encircling the four sides, they made short work of the blobs of lard that dotted the granite, easily wiping them clear with an ever-available supply of T-shirts.

The participation of women, assisted rather than obstructed by their male classmates, was evident. They were being pushed and pulled *up*, rather than pulled down; their muscles as strong and untiring as the men's, their bodies often more lean and lithe, and their athleticism unquestioned, they were ideal candidates for the upper layers, as the entire plebe class worked to ascend to Herndon's apex.

The Dixie Cup was no longer in play, having been dislodged by a well-aimed, greasy wad of torn T-shirt just a few minutes earlier. The cannon boomed again at the Climb's one-hour mark.

Like the Herndon Climb, which has no official time limit, a game of baseball can theoretically last forever. Baseball fans have seen it and the players have played it: when a game is tied at the end of nine innings, extra innings are played until the tie is broken. If the home team, which bats in the bottom half of each inning, scores a tie-breaking run in the bottom of the ninth, or any extra inning, the game is over. So a baseball game can end on a single pitch or a single swing of the bat if the batter hits a home run or drives in a base runner.

At Herndon, Christian Schwien came up to bat for the class of 2022.

He seemed to come out of nowhere, and suddenly he was standing at the top of the pile, on a very firmly established second layer of bodies. That layer was built with an unusually high degree of stability, with an inner and an outer ring of tightly packed classmates, including several women.

A hush briefly fell over the crowd as the tall, slender native of Fairfax, Virginia, stood high, holding a combination cover. Herndon's apex was still about eighteen inches above his outstretched hand when the hush turned to cheers and shouts of encouragement, as Schwien took a shot, tossing the cap gently, very nearly sticking it at the top of the monument. It came off, and he lost his balance, but the strong double-ring of classmates helped him right back up.

Another toss, with the same cap, missed again, and another after that. Each time, Christian was able to catch it with one hand and steady himself for another throw.

On his next attempt, the cap landed neatly, and stayed in place.

Did he get away with a bit of a cheat? Maybe so. In past Climbs, the unofficial rule that the combo cover must be *placed* at the top, not *thrown*, has been more strictly enforced. But such interpretations are entirely within the discretion of the superintendent, and when Christian's last toss stuck at the top of the monument, many spectators looked to Admiral Carter, half expecting him to waive off the attempt and compel the plebes to continue.

Carter was apparently in a forgiving mood, and he allowed the result to stand. The Herndon Climb, the fifth and last of his tenure as superintendent, was over, and the traditional chorus of "Plebes no more!" commenced loudly.

The official, very respectable time was one hour and five minutes. It was the quickest time for a greased Herndon Monument since 1988.

"It was crazy. At one point I fell and hurt my knee a little bit, so I took a breather," Christian said a short time later, facing a bank of reporters' microphones in front of Zimmerman Bandstand as his classmates rejoiced. "And then I saw an opening, and people were throwing me up there, because I'm tall and skinny and they pumped me up. I just threw the cover a couple of times and I finally got it."

As to whether he has designs on eventually reaching flag rank in his Navy career, fulfilling the never-fulfilled promise of a Herndon scaler, Christian was noncommittal.

"We'll see about that. I've got a while to make it," he laughed.

In the Herndon Climb, he certainly demonstrated a rare combination of physical skills, highly prized in the Navy, when he deftly tossed a combination cover and hit the target perfectly, with only a few practice tries while standing on a footing as unsteady as the deck of a storm-tossed ship at sea.

If the guy who runs the ring toss game on a carnival midway ever sees Christian Schwien coming, he'd better close up shop in a hurry or risk losing his entire inventory of stuffed animals.

Newly minted third class midshipmen Brogan Welch and Anna Tenhoopen were satisfied to have the Herndon Climb and their plebe years at the Academy behind them.

After a summer training cruise from San Diego to Victoria, British Columbia, Brogan recalled the anticipation he felt just before the cannon sounded at the start of the Climb.

"Standing in line was a lot more emotional than I thought it would be," he admitted. "Not just overwhelming, but you're just [standing] there, and this is literally it. It's hard to contain all that stuff."

At Herndon, Brogan and his classmate Kyle Lennon found themselves engaged primarily in heavy-lifting duty. "We just kept lifting people up, because we were sort of the bigger guys, so there was no chance we would make it even close to the top. But we started this thing where we'd just pick people up and throw them up, and hopefully get them to the top."

He had just one word to describe the immediate aftermath, when the combo cover reached the top of the monument and plebe year had officially ended for the class of '22: "Insanity."

Anna will always remember the roar of the crowd. The end of plebe year sounded like nothing she'd ever heard or will ever hear again.

"It was the loudest I've ever heard anyone," she recalls, "louder than any rock concert I've been to. The feeling was like a roaring in your chest that you could feel. It was such a relief. The best part was that feeling that the second [the combo cover] touched the top, that we had succeeded so quickly. A lot of my classmates were crying, even the ones that usually wouldn't cry. That feeling at the end was definitely the best part, thinking back on all the stuff we had to do all year, and that finally it was over."

Both of Anna's parents are Academy grads; her father was class of '87 and her mother was '94. So, although she never personally witnessed a Climb, her parents gave her an idea of what to expect. As she recalls their firsthand descriptions from their own experience, the words "sweat" and "sweaty" were mentioned frequently.

"They told me that it was a really sweaty day," Anna laughs, "that it was really gross, and that it would take a few hours, but that when it was over, it was the best feeling to go back to Bancroft Hall and get to walk on the side of the P-way and not have to plebe our covers. That feeling going back into the hall was the best. The anticipation of it was amazing. Thankfully it only took us one hour. But it was sweaty. It was gross. Grass was everywhere. Mud was everywhere. So many sweaty bodies around me at once. It was crazy. But then the feeling going back in the hall was amazing."

Rather than stand on the sidelines, Anna chose to experience the Herndon Climb in all its damp and pungent glory, right among the sweaty bodies.

"I was three, maybe four people out from the bottom of Herndon. I was helping people that were stepping on my head and shoulders," Anna explains. "We were lifting them up and pushing from the sides. Everyone would start yelling, 'Push! Push!' and we would all put our weight forward so that the people slipping down the monument would get pushed back up. I was very much in the middle of it, but I wasn't going up. I wasn't on the edge, either. I wanted to contribute."

"The best part about Herndon was just being there with my buddies," Brogan adds. "There's just something about giving all your physical exertion and diving into that pile with a bunch of your closest friends that's unreal. I suppose it's the same feeling you get when you perform really well in a sporting event. Like Anna was saying, it's awesome. And it really builds your camaraderie. Because you've all done this entire year together and now it's the end."

And the worst part?

"People thinking that they were light enough to climb on top," Brogan says with a laugh. "And you couldn't be mean to them, and say no, you're literally going to fall, don't make me lift you up there. So we'd lift them up, and they'd fall, and then they'd come right back to us and say, all right, lift me up again! That was probably the worst part. But it wasn't even that bad. It was actually comical in a way. But Herndon will always be one of those stories that if you knew the person when you went to the Academy or during the Climb, you will always be able to sit down and talk to them about it."

Afterword

THE ACCIDENTAL ADMIRAL

(or, The Wrong Way to Cut a Green Onion)

> Say the wise: How may I know their purpose?
> Then acts without wherefore or why.
> Stays the fool but one moment to question,
> And the chance of his life passes by.
>
> —The 11th Law of the Navy

Note: Rear Adm. Jim McNeal and coauthor Scott Tomasheski had a series of conversations in the summer of 2019. Those conversations, edited for content and clarity, are transcribed here.

ST: Do you remember the very first time we met?

JM: No. Do you?

ST: No.

JM: Had to be seventh grade. When school started, in the fall of 1975.

ST: Probably in ISP.

JM: Yes! The "Independent Study Program." For gifted kids.

ST: Gifted? How'd we get in?

JM: Good question.

ST: I goofed off more in ISP classes than in regular classes. Way more.

JM: Me too.

(*Laughter*)

ST: Well, we became friends. Survived lots of adventures all through high school and stayed in touch all these years. I think a lot of your friends, myself included, were surprised when you announced you were applying to the Academy. And eventually got in, after NAPS. How'd you arrive at that choice?

JM: Well, my father is a Naval Academy graduate from the class of 1962, and after graduation he went into the Marine Corps and served two tours in Vietnam. A real hero in my book. And as most former Marines do, they run their families in a certain way. That inimitable USMC style. I've always been incredibly independent, not much of a rule follower, two things that were in direct contrast to my dad's parenting style. Anyway, in our junior year of high school, I was told that I'd be limited in my college choices. Financial concerns, you know, so it had to be a commutable local school. UCLA or Cal State Northridge, basically. Neither of which were an attractive option for me. I was ready to leave home, and my parents were ready for me to go.

ST: I bet. So what was your plan?

JM: I didn't really have one, actually. But when I took the SAT at the end of our junior year, completely on a whim, after I checked UCLA and Northridge for my scores to be sent to, I also checked USNA. Later I told my parents about the test, and I casually mentioned that I had sent my scores to the Naval Academy.

ST: As I recall, your father questioned whether you had the right stuff to make it at the Academy. Maybe he inspired you a little bit.

JM: Maybe more than a little bit. Like I said, he's a hero in my book. He was a Marine Corps field artillery officer, and in Vietnam he was a forward observer, or FO, meaning he spent a lot of time in a small airplane flying around and calling in artillery strikes.

ST: That sounds dangerous.

JM: Very dangerous, but fortunately he came away unscathed. He was there most of 1965 and 1966, and he missed the birth of both of my younger brothers. And he really loved the Corps, but at the five-year mark of his service and with three kids under the age of five, it was time to get out. I think he would have also gone into the Reserves, but at that time the Reserves were going to Vietnam too. After two nine-month tours, if you come out unscathed, you don't want to tempt fate and go back again.

ST: A common friend of ours related a story to me, just recently, about something you told him at the time, way back when. That your dad called you out in the middle of a family dinner because you did a lousy job cutting an onion.

JM: Technically speaking, it was a scallion. Or the type of onion with the hollow, tubular green leaves growing directly from the bulb.

ST: The infamous "Green Onion Story"! Please tell.

JM: Well, that particular summer evening we were having tacos for dinner, a family favorite. I was helping my mom, and my job was to help with one of the ingredients, finely chopped scallions. I was never very interested in doing a good job if I wasn't interested in doing it in the first place. Something I passed on to my youngest son. This night's dinner prep was no exception. Instead of cutting all the way through the scallions, I only cut part of the way, resulting in long pieces with lots of little indentations.

ST: Not finely chopped.

JM: No, not at all. So when it came time for my dad to add some to his taco he pulled out the multi-indented scallions, several inches long, and said, "I wouldn't bother applying to the Naval Academy, they would never take someone like you." If that's how I cut a green onion.

ST: Well, he certainly had a point, but what a way of saying it. I'd have burst into tears. But not you.

JM: No. I wasn't hurt by the comment. In fact, based on that type of work product he was probably correct. Actually, the emotion I had was irritation. Mixed with a dash of anger. So I was bound and determined to prove him wrong. In April of 1981 I got my letter, saying I had been accepted as a midshipman candidate at the Naval Academy Preparatory School, and I was overjoyed. I was leaving. And if I could get through a year at NAPS in Rhode Island, well, the Naval Academy *would* take someone like me.

ST: What was NAPS like?

JM: Coldest year of my life.

ST: Sure. Because you grew up in warmer climes. But aside from the weather?

JM: Just for some background info, NAPS is the third oldest school of the Navy, after the Academy itself and the Naval Post-Graduate School. It was started to give an opportunity to enlisted sailors and marines from the Fleet who wanted to go to the Academy. To give them a year to get back into the swing of things academically, because they would have been out of high school for a while. After a period of time, it was used to help other groups succeed, essentially anyone the Academy thought would benefit from an extra year after high school, mostly athletes. And then after that was done, if they had any extra slots, they'd say, are there any unathletic males that we can send to NAPS? (*Laughs*)

ST: Like you?

JM: Yes. And for me it was a maturing year. Because don't forget, my birthday is in November, I was always one of the youngest in our peer group. You spend three weeks, like a mini version of plebe summer, learning the military way of life. You're trained by midshipmen. But then the mids go back to the Academy, and then it's just self-run. Best I could figure out was that I was there because my math SAT score wasn't high enough. And I remember everyone being surprised on the first day of classes, the teachers just started teaching, right out of the gate. Not just taking roll and dealing with people coming in and out, getting textbooks, goofing off for a day or two before really getting started. My first class was trigonometry, with Professor Prassl, and as soon as the bell rang and the section leader reported all present and accounted for, we sit down, and Professor Prassl says the syllabus is face down on your desk, and he starts actually writing stuff on the board and teaching! Everyone was a little stunned.

ST: It's basically the same thing in regular college, like I did. That first year is definitely a maturing year, and yes, they generally start teaching right away. First day of biology lab, there was no cutting out early to goof off and read the syllabus. We were dissecting fetal pigs on day one. It put a lot of people off biology! The class size was a lot smaller in the second meeting.

JM: I'm not surprised.

ST: So what was the transition like after you successfully completed the year at NAPS and started at the Academy?

JM: My fellow NAPSters and I showed up in Annapolis on July 6th, 1982. We had a "private" Induction Day. Not sure of the rationale for that, but we got all of our gear, got our shots, and got haircuts even though we all already had regulation haircuts. This was all accomplished a day before the rest of our classmates would show up.

ST: The "high school kids"?

JM: Right. So, our I-Day experience was simply hanging out in our rooms for most of the day. Around three or four in the afternoon we were told to muster in Tecumseh Court for the swearing-in ceremony.

ST: Were your parents there to witness your official, triumphant entry into the Academy?

JM: No. They were planning a trip to Annapolis in September for my dad's twenty-year class reunion. And two trips in three months was a little too much for the family budget. Anyway, I knew my way around the Academy fairly well, having been down there when I was at NAPS. While most of my classmates hugged their friends and families and then entered Bancroft Hall through the main entrance, I took a circuitous route to my room. Up a back stairway.

ST: So you wouldn't be noticed?

JM: So I could avoid everything. I had figured, correctly it turned out, that our quote-unquote friends in the class of 1983, the seniors, would be lying in wait for the brand-new plebes, who up until the moment they raised their right hand and swore an oath "to protect and defend the Constitution of the United States against all enemies, foreign and domestic," were off-limits.

ST: So they were waiting to give you all a friendly welcome? Personal greeting?

JM: Let's say deeply personal, and not at all friendly. Demanding answers to questions the plebes couldn't possibly answer. Criticizing every aspect of our uniform, the way we spoke or saluted, and so on. Giving us a real taste of what plebe year was going to hold for us. Showing us that things aren't fair, and they aren't supposed to be fair. You just have to take it and make the best of it.

ST: And you avoided all of that by sneaking in the back entrance.

JM: Yes. I get to my room and I'm there and I have two roommates, both straight out of high school, and I didn't see them for a long time. They were just getting destroyed out there. Finally they arrived in our room. Bedraggled and melancholy, thinking of the task ahead of us for the next six weeks.

ST: And if I'm not mistaken, one of your classmates was Peggy Mansfield, who you would eventually marry, although you didn't know her yet.

JM: Yes, Peggy had the opposite experience. Her parents were there and she raised her hand and took the oath and all that stuff, and then she thinks, "Okay! I guess I'll just walk straight into Bancroft!" (*Laughs*) It took her about two hours to get to her room.

ST: (*Laughs*) And you've said she teases you, calls you "the dumb one who had to go to NAPS" because she went into the Academy straight out of high school. Who was the dummy that day?

JM: (*Laughs*) No comment! But having spent a year at NAPS, and sort of having the military piece down, those thirty-nine days of plebe summer were not terribly difficult for me. Marching, saluting, wearing a uniform, and all that had become second nature. Plus I've been blessed with a great memory, so the endless memorization we were required to do was never all that difficult.

ST: Like memorizing phrases from *Reef Points*? Daily menus?

JM: And much more. But the nice thing about the summer was that the ratio worked in our favor, a roughly six-to-one ratio of plebes to first class midshipmen. All that changed when all the upperclassmen returned in August and the brigade reformed.

ST: How were you organized within your company?

JM: My classmates and I, we were dispersed into our company squads. Three squads per platoon, and a total of three platoons in our company. Each squad had a few first class, that's the seniors. And three second class, three third class, and, in our squad, three plebes.

ST: You made it through plebe summer, but you knew that greater challenges were ahead. How ready did you feel?

JM: I wouldn't exactly say I went into plebe year with a great deal of momentum, but I didn't feel that I wasn't up to the challenge, either. When we briefly visited the Academy as NAPSters the previous year, I got a real taste of what being a plebe was about, and I was ready. What I wasn't ready for were two squad mates who struggled mightily with the most basic plebe rates. And a particular senior who let's say took a very strong interest in my professional development.

ST: You had a nemesis!

JM: Oh, yes.

ST: Let's get to him later. What about the squad mates? Did they hold you back? And what exactly are "rates"?

JM: Rates in this context means all the information a plebe is required to know on a daily basis. And, for whatever reason, my squad mates couldn't figure it out. You see, I'm a student of leadership. I taught a college course in leadership. And one of my very earliest leadership lessons I learned was from my squad leader, first semester. When I knew my plebe rates and my squad mates didn't, rather than chastising them and holding them to the standard, he got furious with me. Holding me responsible, in other words, for their lack of knowledge.

ST: For example?

JM: Well, there was one piece of information we were required to recite on demand. The days, which were four numbers: the number of days until the Army-Navy football game, days until Christmas leave, days until the Second Class Ring Dance, and of course, until graduation. So the "days" were four specific numbers that each shrunk by one, day after day. And it gave my squad mates fits. We were required to eat with our squads every lunch and dinner, starting Sunday evening through Friday lunch. During these meals, eating was "optional" as we were constantly tasked with reciting whatever bit of plebe knowledge we were responsible for knowing. And while the upper class quizzed us endlessly on plebe rates, we had to fill their glasses, serve their food. Family style. Quite a multitasking effort.

ST: Maybe that was the point.

JM: I believe it was. Anyway, there was one disastrous attempt at the days by my peers during an evening meal. I was told in no uncertain terms that they had better know them cold at the following day's noon meal. Next day, when we walked into King Hall for lunch, I made sure they knew the right numbers, confirmed they had them down. I was relatively confident. Surely they could regurgitate the numbers on command in the next ten minutes?

ST: Let me guess. They both blew it.

JM: Yep. Neither one could come up with the right numbers. So I was asked to recite them, which put me in an awkward position. I asked permission to not "bilge" my classmates, a concept that had been drilled into us since Induction Day. But by doing that, I was telling them I knew the answer but didn't want to show up my classmates. The conversation quickly turned into my defiance of the previous day's order that they both know the answer and be ready with the correct days. It seemed to further cement in my squad leader's mind, McNeal was not a team player.

ST: Sounds like you were in a no-win situation.

JM: Pretty much. Going into final exams I got my first semester evaluation from my squad leader. This was the opening sentence: "The only things that Midshipman Fourth Class McNeal excels in are Academics, Athletics, and Plebe Rates." The rest went into detail about how I wasn't a team player because my peers did not know their required knowledge. Justifying my class ranking in my company, a lofty one from the bottom.

ST: John McCain was fifth from the bottom in his graduating class. It's the name of a chapter in one of his books. "Fifth From the Bottom."

JM: I'm in good company, I guess!

ST: So this was your first leadership lesson?

JM: In this case, the lesson was what *not* to do as a leader. But that stinging rebuke at the very end of my first semester at the Academy was softened a bit because I was heading off to California for Christmas leave and almost three weeks away from the Academy.

ST: To hang out with yours truly and your other old pals. Possibly drink a few well-earned beers.

JM: No doubt. But all the while, in the back of my head, I knew I would have to get on an airplane and come back to Annapolis in early January, which provoked a sense of dread.

ST: Why?

JM: Spring break was eight long weeks away. Coming back from Southern California into the cold and bleak and dreary winter days, known as the "Dark Ages." They weren't fun for any midshipmen, let alone plebes. I knew at least, we would get new squad assignments for second semester. I was going to get a fresh start. Or so I thought. My first semester squad had one other firstie, and he was a real joy.

ST: Ah. Your nemesis.

JM: He was an athlete, and although he rated sitting at a team table, he chose to eat with our squad. He was always there for noon meal, but for evening meal he always arrived late from practice. Now, I should point out that while we were quizzed incessantly while also serving food and drink to the upper class, we did get a chance to eat after the quizzing was done. At lunch, it depended on the length of our little "chats" and our class schedule, but we might get to gulp down a few bites. More often than not we left the table still hungry. Unsatiated. Evening meal was a different story. We could stay to eat for as long as we wanted until they cleaned up King Hall. No other commitments but a stack of schoolbooks waiting for us. We ate in the required plebe fashion: one bite at a time, sitting on the front six inches of the chair, eyes "in the boat" or straight ahead. At least we were eating. But "the Athlete" always forced us into a particular Sophie's Choice. Do we leave as soon as the last upperclassman leaves the table, sacrificing dinner, or do we risk a few bites of food before the Athlete shows up? Knowing that if he beats us to the table, he won't let us leave or eat, and we'll spend the next twenty minutes or so getting quizzed again as all of the food was taken away?

ST: Tough call.

JM: Sometimes the decision was made for us. But I'll never forget the feeling of trying to steal a few bites of food and get the heck out of there, when I would see in my peripheral vision a glimpse of those long strides. Those long, loping strides. The Athlete, as he sauntered down King Hall. Sent a chill down my spine.

ST: And you got along even better, second semester.

JM: I got back to the Academy on a Sunday night and saw our new squad assignments. Guess who was my squad leader? Yep, the Athlete. As we formed up for the first time he cackled. Delightedly. "Oh, Mr. McNeal! I didn't know *you* were in my squad!" My heart sank. And he never actually said so, but I feel sure that his intent was for me to leave the Academy. To punch out or bilge out. And he would have been delighted if I had. Of course, for me, going home was not an option, in any way, shape, or form.

ST: So you were stuck together.

JM: We were stuck together. He seemed to take particular delight in tormenting me, as did his roommate, who was on a battalion staff previously and was now in our company area.

ST: Example?

JM: One lunch was a great example. I should explain, one of a plebe's duties is to perform what are called chow calls. At five and ten minutes before noon and evening formation are chow calls. They are ostensibly to let the upper class know they have to get ready for formation. But the true purpose is another memorization drill. Knowing the number of minutes until formation, the officers of the day, officers of the watch, the menu for that meal, the uniform required for the formation, sports that day. All sorts of information, screamed at the top of your lungs, and as rapidly as possible. Stumble at all and the upper class pounces, like jackals. Things get ugly very quickly. Fortunately, with my good memory, chow calls were easy for me. Even when I had to go mano a mano, so to speak, with my second semester squad leader.

ST: With the Athlete.

JM: One day the Athlete ordered me to come to his room and give him a set of personal chow calls for noon meal. Not only at the ten- and five-minute marks, but at ten, nine, eight, seven, six, and so on, all the way to the one-minute mark. Now, most first class mids have lots of disposable income relative to the other classes, and most had nice stereo systems, which was a privilege they rated. The Athlete had a very nice, very loud stereo system. This particular day, it was playing softly when I got to his room. He told me to come in and shut the door, and at that moment I knew something was up, and it wouldn't be good for me. So formation was at 12:10. As the clock ticked to 12:00 I launched into this barrage of information. Screamed at the top of my lungs. And as I did, he turned up the volume on the stereo, to a deafening volume.

ST: He cranked it up to eleven?

JM: Yes, and at the same time pantomimes that he can't hear me, that I need to "speak up." By the end I'm literally screaming as loudly as I can, but it was like there were no words coming out of my mouth. When I finished, he turned down the volume on the stereo. That was just the first chow call, and I had to repeat this nine more times. Finally, after the last one, I raced to my spot in formation. I was a little late, and naturally the other upper class loudly chastised me for that.

ST: Unbelievably sadistic. I wonder how he thought of something like that?

JM: Oh, it gets worse. We have a tradition of "hundreds night," which is a hundred nights before graduation. Age-old tradition at the Academy. On that particular evening the fourth class and first class trade places, just for a few hours. So the plebes get to yell and scream at the first class, but knowing that the next day things go back to normal, and any unruly plebe behavior is tempered by the knowledge that there were ninety-nine potential days of reprisals afterward. But I didn't care about reprisals; how could it get any worse? The chance for a little payback on the Athlete, just once, it was like an unquenchable thirst.

ST: He certainly had it coming. Was it satisfying?

JM: Well, I was sitting in his seat at the dinner table, knowing he'd be late as usual. And I was wearing the rank insignia of midshipman ensign, per the custom. When I saw those long, loping strides this time, I didn't feel that usual sense of dread. I felt gleeful. Ready to give him a dose of what he'd subjected me to since August. Well, my first clue about his intentions was that he was still wearing his first class insignia. Dressing the part was a crucial piece of the role reversal, and he wasn't playing. So he approaches the table and I told him to take a seat. And he says, "McNeal, get out of my [expletive] chair."

ST: Oh, that's wrong.

JM: I repeated my order, told him to take a seat, and he repeated his answer. I was exasperated, practically on the verge of tears. "But it's hundreds night!" I told him, but he didn't care. "I don't give a [expletive] about hundreds night and for the last time, get out of my [expletive] chair!" he answered. I was devastated.

ST: One of the other upper class should have called him on that.

JM: They didn't.

ST: Did the reprisals get worse?

JM: No, but they didn't ease up on us, either. They continued on right until the very end, including the very morning of the Herndon Climb. There was some perceived slight at the lunch table and they rebuked a few of my classmates, and me, with "You are still plebes to us and you'd better act like it."

ST: But finally, you made it to the Herndon Climb. How was the weather that day?

JM: Hot and humid. Typical Annapolis presummer weather. But that was the last thing on our minds. Gathering in T-court, we knew plebe year was just that close to being over. When the cannon sounded and we sprinted out to Herndon, I settled in at the base of the monument.

ST: To be at the bottom? And have everyone standing on you?

JM: Yes. And I couldn't have been happier. Immediately drenched in sweat, but settled in, not knowing what was happening above me. I heard the crowd ooh and ahh, then groan, when a body tumbled down. . . . I hoped we were making progress, but it really didn't matter. I was with my classmates, working toward the common group goal. And I was still standing despite every effort of the Athlete. The toughest year of my life was in my rearview mirror. And at the one hour, forty-three minutes, and fifty-five seconds mark—

ST: You remember the exact time?

JM: No, I looked it up recently. But Kevin Delamer was the guy, he got it for our class. Pulled down the Dixie Cup and put on the combo cover. And that was all. Finally, plebe year was over.

ST: And how did that feel?

JM: It was a combination of pure joy and emotional release. When I walked back to my room in Bancroft, this time I did exactly the opposite of what I did on Induction Day after swearing in. Instead of sneaking up a back stairway, I went through the main entrance into the rotunda, taking my sweet time. I made sure I was in Bancroft for as long as possible. I just sauntered along, no chopping, no squaring corners, just sauntering along, just because I could do it now. Just savoring every second of it. One thing I'll never forget is when I got back to my room, I turned on my formerly contraband boom box, which was now legal. The first song I heard was U2's "New Year's Day" blaring into the room. Yes. It felt like a new year. Hearing that song will always make me think of that moment.

ST: You once told me that the Herndon Climb was the second greatest moment of your life. Do you stand by that?

JM: Second only to Graduation Day, three years later, throwing my cover in the air at the stadium.

ST: So after you guys completed the Climb, in one hour and . . . how many minutes was it?

JM: One hour, forty-three minutes, and fifty-five seconds.

ST: Okay. So you were no longer a plebe. Did the Athlete congratulate you and shake your hand?

JM: No. By then the seniors were getting ready to graduate, and I didn't see him the rest of my time at the Academy. But I did run into him, quite by chance, one other time. I was almost four years out of the Academy, I even remember the exact date, May 1st, 1990, because that was the day I departed my second ship, the USS *Merrill*, in Pearl Harbor, on our way back from a six-month WESTPAC deployment. I headed to the Honolulu airport, resplendent in my summer white uniform as a Navy lieutenant. Flying home to San Diego to see my wife, Peggy, and my now eight-month-old daughter. I was quite giddy. So at the airport I settled into a seat by the gate and an older couple sitting across from me struck up a conversation.

ST: Civilians love to strike up conversations with uniformed service members, at least in my experience.

JM: Yes they do, and we are honored to speak with them. So we chatted away for a few minutes until, out of the corner of my eye, I saw something that sent a chill down my spine. Brought back emotions I hadn't felt in almost seven years.

ST: Oh, no . . .

JM: That long, loping stride I knew so well. Kind of lumbering, swaggering. Could it possibly be? I stopped talking midsentence, and the lady across from me asks, "Have you seen a ghost?" Not quite a ghost, I said. I asked them to please watch my bag and I took off in hot pursuit. I walked quickly down the concourse a few gates and there he was. Also a lieutenant in his summer whites. The Athlete. I walked up and stood right in front of him. There was no emotion in his eyes and I realized he had no idea who I was. Finally there was a glint of recognition, and he stammers, "Hey, it's Jim, right?" I answered yes, and I added, "I can't believe after all you did to me, you don't remember me." After more stammering he managed to get out, "Well, how are you?" I didn't know what to say. I simply stated, "Despite your best efforts I turned out okay." Not sure what I meant by that but the conversation ended as I turned around and went back to my gate.

ST: And that's the last time you ever saw that long, loping stride?

JM: Yes.

ST: Let's talk about sports. I understand that all midshipmen, including plebes, are required to participate in a sport. Varsity, club sports, or intramurals. First of all, how does anyone have time to practice and play sports with everything else on their plates?

JM: Part of the Naval Academy mission with the midshipmen is to develop them morally, mentally, and physically. Thirty or forty percent of midshipmen are involved in varsity sports, but everybody has to be in a sport, so if you're not in a varsity sport or a club sport, then you have to do intramurals. The time is blocked out, from the last class period until dinner. They actually build it into the schedule.

ST: A varsity sport is, for example, football or basketball or other sport that competes in NCAA-sanctioned games?

JM: And a club sport is a club. Literally a club. So you have to pay dues, you have a club president. There are lots of club sports, every school has them. They play against each other at the club level. For example, one of the newest club sports at the Academy is women's softball. They won the national championship two years in a row at the club level. Of course, they're not playing Oklahoma or UCLA or any of the NCAA teams, but there are many other schools that have very good club teams. I played rugby, which was a club sport. We did pretty well, we just played a lot of local teams, because we didn't have a travel budget. I don't remember our exact record, but we did pretty well.

ST: How did you get involved in rugby? We didn't have a rugby team in high school.

JM: Well, we came down from NAPS in the fall of 1981, and we were embedded with some plebes, to get a taste of it. I saw how brutal it was. Particularly in King Hall, with the quizzing on rates, sitting on the edge of your chair, all of that. I thought, I've got to figure a way out of this thing.

ST: How?

JM: I noticed that there were team tables, where the varsity athletes sat with their teams, where they eased up on all that stuff. I came up with this grand idea that I'd do track, because the track team was a varsity team. At NAPS we had a track team, we played other prep schools like Exeter Academy and Phillips Academy. We didn't have any coaching, and I was trying to throw the shot put, and I just didn't know what I was doing. I showed up for plebe summer and said, hey, I'm a track guy, hoping to get on the track team at least through plebe year, so I could get on the team tables. Well, I went out in the fall and I realized I was in way over my head. You can't just have never done something then do it at the college level. There was no way it would happen. So I ended up in intramurals for the rest of plebe year and for the first semester of my youngster year. Second semester, my roommate and I were split up by our company officer because we ended up, as he said, "being negative influences on each other."

ST: You were a couple of screw-ups.

JM: Let's just say we liked to have fun, regardless of Academy regulations. He thought he could squash that by splitting us up, so I moved in with a couple of other guys. Across the hall from us were three guys in class of '85, so second class, and they all played rugby. I became friends with them, and they said hey, why don't you come out and play. And so I went out and played and just loved it. That's how I got into rugby. One of the nice things about rugby, at least back then, is that it was a level playing field for everybody. I think we had a couple kids who played in high school, but for the most part it was an unfamiliar game to everyone.

ST: One thing I know is that it's a rough game. You had a couple of injuries that ended up influencing your career path.

JM: They have intramural rugby and club rugby, and they encouraged the newer players to play both to get experience in the game. I was playing intramural rugby when I dislocated my right shoulder. I returned for my junior year and my right shoulder was relatively okay. Not great, but okay. But then I dislocated my *left* shoulder. I rehabbed it, and it healed, mostly. The medical staff made me a brace, and I wore that and actually I was okay. I came back for my senior year, and maybe I didn't rehab well enough or I didn't keep my strength up or whatever, but within a couple of weeks, almost every practice both shoulders were coming out. I just finally said, this isn't going to work. So I had to stop playing during senior year, and that was disappointing. But the good news was that because of my shoulder problems I was able to go to Supply

Corps, which was what I really wanted and it ended up being the best career path for me.

ST: You couldn't do anything extremely physical?

JM: At the end of junior year, everyone has a complete physical exam. They process all the results, and then you meet with a doctor at the start of senior year to go over them. When I met with the doctor, he says, wow, you've got some messed up shoulders.

ST: Isn't it great when a doctor tells you what you already know?

JM: Well, he gave me a couple of choices. I should explain, in the Navy there are various communities to choose to serve with. One is the unrestricted line community, which is Aviation, Surface Warfare, Submarines, Special Warfare, and, of course, the Marine Corps as well.

ST: Highly demanding, physical activities.

JM: Yes. There are also restricted line communities and staff communities. The doctor said if I really want to go unrestricted line, I'd need surgery on both shoulders. This is long before the days of arthroscopic surgery, so they'd be painful and very invasive surgeries. Very long, very arduous rehab. And even if you're injured, you have to still pass all the physical tests before graduation. He said, candidly, it was going to be a real race to the finish line for me, having to deal with both shoulders. That's option A. Option B was restricted line, and I said I was interested in the Supply Corps, which is restricted. It's the business side of the Navy. So it was easy, he just checked the box for restricted line and I was all set. I was happy.

ST: And all these years later, you still play a little rugby. The "over fifty" games against Army alumni. Personally, I think you are out of your mind. Guys over fifty play softball. Maybe basketball. But rugby?

JM: I've played two years in a row. I actually wore a brace this year on my right shoulder, which is the worse of the two. Last year I just winged it, but this year I have a brace, so my right shoulder is pretty well tied up.

ST: What are the games like? Sort of a war of attrition, is what I'm imagining.

JM: The first year we played, in 2018, we had twenty-five guys at our practice the day before the game. It was really just a meet and greet, and everybody just touching a rugby ball for the first time in a long while. Rugby has fifteen players per side, so we had fifteen on the field and ten substitutes. Immediately, as soon as the game started, people were dropping like flies. Pulled hamstrings, sprained ankles, etc. We ended up with three guys out of the twenty-five needing surgery due to injuries. I came in the game five minutes in, as one of the first subs, and I ended up playing the entire game. At one point one of my teammates was just so tired and he looked over and said he needed a sub. I said, look at that sideline! Do you see any subs over there? We had guys with icepacks, crutches. Guys bleeding.

ST: (*Laughs*) Like the war wounded. Like an army field hospital after a battle.

JM: Yes, exactly.

ST: Well, you guys are tough, that's for sure. And you'll be playing again next year?

JM: Definitely. They're talking about having our game, the "alumni" game, be right before the real Army-Navy rugby game, with the current midshipmen and cadets.

ST: That sounds like a lot of fun. And of course your desire to "beat Army" is as great as ever.

JM: Oh, of course.

ST: It sounds like you're enjoying your life after retiring from a long career in the Navy. Speaking of that career, you eventually earned an admiral's star and a job as deputy commander and then commander of NAVSUP Global Logistics Support. I'm looking at your curriculum vitae: GLS was an Echelon III command responsible for eight Fleet Logistics Centers, 149 sites, almost five thousand civilians, and two thousand military personnel. One-billion-dollar budget. That's "billion" with a "B." You've been to many different countries, military bases all over the world, including hot spots like Iraq and Afghanistan, a real lifetime of adventures. What made you want to write a book about the Herndon Climb?

JM: One of my classmates, also a former NAPSter, Dan Ballister, has a simple one-sentence line in the *Lucky Bag* yearbook, describing our four years: "Wasn't easy, wasn't supposed to be." As tough as plebe year was, and as the four years at the Academy were overall, I wouldn't be the person I am today, have the family I have today, have the career I've had, without the Naval Academy. I truly owe everything in my life to my time at USNA. The Herndon book highlights a lot of stories from other graduates who've been similarly affected by their experiences at the Academy. I hope that my fellow alumni read the book and think about their time at the Academy and reflect on all of their experiences and memories of their time as midshipmen. The Herndon Climb meant so much to me, and I hope anyone reading the book will understand why it is such a wonderful and unique tradition.

REFERENCES

Banning, Kendall. *Annapolis Today.* Revised by A. Stuart Pitt. Annapolis, MD: Naval Institute Press, 1938–63.

Gelfand, H. Michael. *Sea Change at Annapolis: The United States Naval Academy, 1949–2000.* Chapel Hill: University of North Carolina Press, 2006.

Hancock, Joy Bright. *Lady In the Navy.* Annapolis, MD: Naval Institute Press, 1972.

Herndon, William Lewis, and Gary Kinder, eds. *Exploration of the Valley of the Amazon, 1851–1852.* New York: Grove Press edition, 2000.

Kinder, Gary. *Ship of Gold in the Deep Blue Sea.* New York: Vintage Books, 1999.

Lucky Bag. Annual yearbook of the U.S. Naval Academy. Various editions.

McCain, John, with Mark Salter. *Faith of My Fathers.* New York: Random House, 1999.

Mines, Janie L. *No Coincidences: Reflections of the First Black Female Graduate of the United States Naval Academy.* Fort Mill, SC: Custom Messages, Inc., 2018.

Reef Points: The Annual Handbook of the Brigade of Midshipmen. Annapolis, MD: Naval Institute Press, various editions.

Shine, An. *A USNA Mom's Journal: Tips and Wisdom from Our Family's Plebe Year through Commissioning and Beyond*. Jang-X Publishing, 2019.

Staubach, Roger, with Frank Luksa. *Time Enough to Win*. Word Publishing, Inc., 1980.

Thompson, Tommy, with Columbus-America Discovery Group. *America's Lost Treasure: A Pictorial Chronicle of the Sinking and Recovery of the United States Mail Steamship* Central America. New York: Atlantic Monthly Press, 1998.

ACKNOWLEDGMENTS

I WOULD LIKE TO THANK EVERYONE who talked to us and shared with us their personal stories about the Herndon Climb, including: Ed Linz, Rob Ballister, Randy Stefanovic, Catherine Ripley, James Nail, Tom Lindblad, Rex Trudell, Stanley Bryant, Brian Weaver, Kevin Scott, Ryan Dupnick, Joshua Caleb Williams, Kevin Delamer, Jeff Bravo, Mark Harper, Jason Cronin, Joshua Dale Gillespie, Shannon Murray, Nelson Santini, Pamela Pitkin, Jerry Witowski, Steve Fortmann, Meagan Flannigan, Karen Halverson, Nicole Battaglia, Barbara Ives, Harrison Smith, Dan Gibbons, Jim Golladay, Dave Kennedy, Dan Morris, Mike Schwob, Jason Morin, Aaron Rosa, Cate Sheerin, Greg Reichel, Shane Halloran, Chris Reaghard, Dave Gordon, Mike Germano, Kelly Laing, Jessie Baynes, Joe McGraw, Anna Tenhoopen, Wesley Nygard, Jadd Oweimrin, and Brogan Welch. Go Navy, beat Army!

Very special thanks and warmest regards to Carrie Dickmann, Doug Dickmann, Kyle Dickmann, Cassie Dickmann, Chris Bianchi, and Rita Barrie Bianchi.

Love and thanks to my wife, Nancy, for her research and transcription skills and her enthusiastic support, expressed in countless ways.

Thanks to Paul Merzlak, Susan Brook, and the staff at the Naval Institute Press for their guidance and hard work.

Vice Adm. Ted Carter, congratulations on a successful tour as superintendent. We wish you fair winds and following seas!

Special thanks to Jim Cheevers, who generously shared a portion of his vast experience with us, and to the staff at the Naval Academy Museum for maintaining one of the world's greatest collections of naval artifacts.

Many thanks to Carmen Farmer, Rosemary Sheridan, Cathy Genson, and Sondra Levy, who helped us learn to write and were unfortunate enough to have both of us in their English classes.

Thanks to Susan Weisberg (and a certain fictional mouse) for helping to inspire us.

A big shoutout to our friends Paul Feinberg, Glenn Sacks, and Steve Spinoglio for their advice, friendship, and patience.

Finally, to my lifelong friend and writing partner Jimmy: Thank you for helping me learn about the Navy, the Academy, and the Herndon Climb. For me, working on this project with you has been the experience of a lifetime.

—S. T.

When Scott called me in May of 2015 and asked if I had ever heard of the Herndon Climb, I had no idea where his idea of writing a book about the Herndon Monument and the Climb would lead. Candidly (and embarrassingly), other than cursory knowledge of Captain Herndon going down with his ship, I didn't know any of the history of William Herndon, the monument, or the Climb; the only Climb I ever was a part of or witnessed was my own in 1983.

Once we dug into the project, we realized that we were going to need a *ton* of help to craft a book worthy of Captain Herndon's legacy. I figured that if I sent out a request for stories from various LinkedIn groups, Facebook groups, and other social media, my shipmates would not disappoint, and they didn't; the response to my posts was overwhelming! Thanks to everyone who chipped in with either a chapter idea or a tidbit of info or an actual interview. This book is for you.

Special thanks are reserved for the Dickmann family, who despite the loss of their beloved Kristen were incredibly gracious with their time, as were the Bianchi family, who suffered a different type of loss but one not any less devastating.

Thanks to Roger Miller, whose photograph of the class of 1998's climb is a valuable addition to the book.

Thanks to Gary Kinder, whose book *Ship of Gold in the Deep Blue Sea* brought Captain Herndon to life and is a fitting testimonial to a great naval officer.

To Lt. Cdr. Terence Preble, USNA '83, who ensured that I had a proper plebe year, RIP.

To my parents, Bob (USNA '62) and Judy McNeal, who provided me with a great upbringing and set me up for success.

With many thanks and undying love to my wife, Dr. Peggy Mansfield McNeal, USNA '86, who always provides me with great advice (when I listen); in this instance, her counsel was invaluable as we explored the Climb from a different perspective.

This book would simply have been a mere idea if not for the Tomasheskis. Nancy provided all our interview transcription, a tedious task, as well as invaluable research. Thanks for all of your hard work; we could not have done it without you. And to Scott, my friend since we met as seventh graders in the fall of 1975, thanks for your hard work, ideas, encouragement, and most of all your friendship. The best part of working together on this project was getting to spend time with you.

—J. M.

ABOUT THE AUTHORS

Rear Adm. James R. McNeal, SC, USN (Ret.), was born in Hawaii and raised in Southern California. The son of a 1962 USNA grad, he graduated from the Naval Academy in 1986. After six years on active duty in the Supply Corps, he transitioned into the Reserve component, retiring in 2017. He is an owner in several businesses, but his passion is coaching with the USNA Sprint football team. He and his wife, Peggy, also a 1986 USNA grad, have three children and a grandson.

Scott Tomasheski is a Los Angeles–based novelist and writer, author of three books in the Time Defenders series. He has worked as a chemist, microbiology laboratory supervisor and manager, and small business owner. He lives in Los Angeles with his wife, Nancy, and a variety of small dogs.

The Naval Institute Press is the book-publishing arm of the U.S. Naval Institute, a private, nonprofit, membership society for sea service professionals and others who share an interest in naval and maritime affairs. Established in 1873 at the U.S. Naval Academy in Annapolis, Maryland, where its offices remain today, the Naval Institute has members worldwide.

Members of the Naval Institute support the education programs of the society and receive the influential monthly magazine *Proceedings* or the colorful bimonthly magazine *Naval History* and discounts on fine nautical prints and on ship and aircraft photos. They also have access to the transcripts of the Institute's Oral History Program and get discounted admission to any of the Institute-sponsored seminars offered around the country.

The Naval Institute's book-publishing program, begun in 1898 with basic guides to naval practices, has broadened its scope to include books of more general interest. Now the Naval Institute Press publishes about seventy titles each year, ranging from how-to books on boating and navigation to battle histories, biographies, ship and aircraft guides, and novels. Institute members receive significant discounts on the Press' more than eight hundred books in print.

Full-time students are eligible for special half-price membership rates. Life memberships are also available.

For a free catalog describing Naval Institute Press books currently available, and for further information about joining the U.S. Naval Institute, please write to:

Member Services
U.S. Naval Institute
291 Wood Road
Annapolis, MD 21402-5034
Telephone: (800) 233-8764
Fax: (410) 571-1703
Web address: www.usni.org